MW01608988

Exploring the Upper Yukon River

Carmacks to Dawson City

by
Gus Karpes

hancock

house

ISBN 0-88839-421-7
Copyright © 1998 Gus Karpes

Cataloging in Publication Data
Karpes, Gus, 1941 -
   Exploring the Upper Yukon River from Carmacks to
   Dawson City

   Includes bibliographical references and index.
   ISBN 0-88839-421-7

   1. Yukon River (Yukon and Alaska)—Guidebooks. 2.
Canoes and canoeing—Yukon River (Yukon and Alaska)—
Guidebooks.
I. Title.
FC4045.Y85K37 1998   917.19'1043   C98-910164-9
F1095.Y9K37

Editor: Nancy Miller
Production: Ingrid Luters
Cover Photo: Gus Karpes

*We acknowledge the financial support of the Government of Canada through the
Book Publishing Industry Development Program for our publishing activities.*

Published simultaneously in Canada and the United States by

HANCOCK HOUSE PUBLISHERS LTD.
19313 Zero Avenue, Surrey, B.C. V4P 1M7
(604) 538-1114 Fax (604) 538-2262

HANCOCK HOUSE PUBLISHERS
1431 Harrison Avenue, Blaine, WA 98230
(604) 538-1114 Fax (604) 538-2262
Web Site: www.hancockhouse.com *email:* sales@hancockhouse.com

Contents

Dedication

To my partner Irene, who continues to travel the trail with me.

Foreword

Whether you arrive in Carmacks by road or by river, you are about to take a trip through some of Canada's most colorful history.

For a time, during that period of history when the eyes and hopes of people from all over the world were focused on the Klondike gold, the Yukon River ranked as one of the most important trade and transportation routes in the world. In the late 1890s, in a period of two years, more than 50,000 people traveled this part of the river to the gold creeks of the Klondike.

Many of them arrived much too late to benefit financially in any way from their endeavor and left the north the same year they arrived. They left just as broke as when they first came, but very much richer for the experience and having participated in what was, in all likelihood, the last great gold rush in the world.

A number of those first stampeders stayed. They dispersed along the river. They stayed, perhaps not as the "Klondike Kings" as they had first envisioned themselves, but as pioneers in a beautiful, but harsh and unforgiving, northern land—a land like no other place on earth. Imagine, if you can, arriving in the middle of the wilderness with nothing more than an ax with which to build yourself a log cabin and a rifle with which to acquire food. Staying and surviving under those conditions was a remarkable feat in itself. Those settlers survived, despite the long, dark and cold winters, with temperatures down to minus seventy Fahrenheit, despite the isolation, the lack of medicines and other social assistance. They not only survived, but also chose to stay and live out their lives here.

There is an abundance of recorded history about Dawson City and the Klondike creeks. Unfortunately, much of the personal history of these stalwart pioneers that lived along the river is lost. Here and there we may find the remains of their cabins and homes. We may follow their trails and fish in their special places. We may find signs of their passing and even share in some of their triumphs and dreams as we gaze into a camp fire on some lonely spit of sand along the river. Sadly, for the most part, any physical record of their accomplishments, dreams and exploits has been reclaimed by nature.

During their short tenure in the land, there were also more than 200 stern-wheeler steamboats traveling the Yukon River. A great number of them toured the very stretch of river that you are about to travel. Many of the boats, built during that heady time of the Klondike gold rush, completed only their maiden voyage. Some were hastily built, slipshod vessels of dubious quality that only managed that one glorious trip before succumbing to bad engineering or pilotage and other hazards. Many of them, at the whim of their owner's fortune, changed hands as much as three times during their maiden voyage to the Klondike. Some were crushed in the ice, some were burned and some were simply beached by their owners as the initial excitement of the gold rush waned. Some of them plied these waters until well into the twentieth century and did not officially retire until the early 1950s. Again, here and there we may find the rotten remains of these once-proud ships, but, by and large, these too have disappeared.

As you launch your boat or canoe into the river at Carmacks, you have the privilege of being able to follow in the footsteps of these early pioneers. You have the opportunity to travel the waterway where once a splendid fleet of stern-wheeler riverboats beat their way into the current. You may be a little better equipped and you may not have to concern yourself with a cold northern winter, but all of the excitement of a journey into raw nature, the exhilarating quest for the Klondike and the adventure of it all has not changed. It is there for you to experience and enjoy to the fullest. Have a great trip.

Gus Karpes, 1997

Introduction

The Yukon River between Whitehorse and Dawson City is split into two guidebooks. These are:

Exploring the Upper Yukon River: Whitehorse to Carmacks
–200 miles (320 km);
Exploring the Upper Yukon River: Carmacks to Dawson City
–260 miles (416 km).

Each of the above segments is generally considered to be a seven-day journey by canoe. The shorter distance of the Whitehorse to Carmacks leg includes a thirty-mile Lake Laberge paddle. The slower going of the lake crossing and the possibility of getting weather-bound make the time lines more-or-less equal

Aside from the distances involved, the two river segments differ in many other ways. The first leg subjects you to an almost pure wilderness experience. You start out in the clear waters of downtown Whitehorse and the Thirty Mile and other than fellow canoeists, you encounter very little else but nature in the raw. At Carmacks, the river is still relatively lively and clear, but already it has gained in size and volume as four major tributaries and a number of smaller creeks have joined in its makeup.

The average cubic meters of water per second at Whitehorse is measured at 240. The average flow at Carmacks is 747 cubic meters per second. It dramatically increases in the second leg and is eventually measured at Dawson City as 2,210 cubic meters. On this leg, seven more major rivers flow into the Yukon River. There are people that live and work along the river, historical places to see and visit. Its character changes from an intimate wilderness stream to a large, commanding and sometimes intimidating waterway. The additional flow of such large tributaries as the Pelly, the White and the Stewart rivers enlarge the Yukon to where, in many instances, it is impossible to see both banks at the same time. The large amount of silt that these tributaries dump into the Yukon creates impromptu islands and bars.

History is more significant in the second leg, with such places as Minto, Fort Selkirk and all of the creeks closer to Dawson City showing signs of past grandeur and commerce. Finally there is an arrival in the Klondike. As Moosehide Mountain and the Dome above Dawson City come into view, you are literally sharing the experience with the more than 30,000 stampeders that landed in the year 1898.

The information I have supplied in this guidebook is accurate to the best of my knowledge. That is not to say that I did not speculate a time or two, but I did so with a certain amount of facts in hand. The beginning of the book gives a brief history of the Yukon River followed by some general wilderness traveling tips and behavior practices that I have found useful over the years. The middle of the book, pages 21 to 48, breaks the 260 miles of river between Carmacks and Dawson City into segments. The length of the segments varies. I have divided the segments at spots on the river where significant change takes place in its character or makeup, i.e., a hazard is encountered, there is a distinctive scenery or water change, there is a landmark, etc. These spots are hard to miss and help to find your place on the accompanying maps; many people find it hard to relate a map to a visual reference around them, so these locations are very helpful.

The next part of the book, pages 49 to 92, provides a historical background to the trip. The numbers appearing ahead of the names on the map are keyed to the numbering in this section. Lastly, pages 93 to 105 include a number of general interest articles and stories that I felt might be of interest and make good reading around the campfire.

In the map pages, the following characters are used:

 shows a high, clay bank borders the river.

 indicates true north.

 indicates the location of an old cabin or building site.

 symbolizes a potential camp site (two or more means it is suitable for a group).

A Short History of the Yukon River

The "Upper" Yukon River is considered to be that portion of the river from its headwaters at Marsh Lake to Dawson City approximately 500 miles downstream. It is in these first 500 miles that its character starts to unfold. Coming out of Marsh Lake it begins as a swiftly flowing, clear-running stream. On its way to Dawson City it slowly grows in size and stature as eleven major tributaries add their flow to its make-up. In these 500 miles it develops into a watercourse that is considered to be the tenth largest in the world.

The documented discovery of the Yukon River spans almost 100 years. Its existence was first mentioned in the journals of Alexander Mackenzie. In 1789, near the end of his voyage down the Mackenzie River, his Native guides made mention of a river on the other side of the mountain range to the west. This river, they said, was of such a size so as to make the river they were traveling on appear insignificant.

The first recorded discovery in 1835 is credited to Andre Glazunov, a trapper and trader working for the Russian American Company. He entered the Lower Yukon from the Anvik River. Glazunov named his discovery the Kwikpak, his interpretation of the name used for the river by the local Natives. Translated from their dialect this meant "Great River."

The next recorded events are more meaningful to us as they deal with the upper part of the river. In the summer of 1843, almost simultaneously, two Hudson's Bay Company traders, Robert Campbell and John Bell entered the Yukon River Valley. Campbell came into the Yukon River at the mouth of the Pelly River. Bell entered the river a considerable distance downstream at the mouth of the Porcupine.

Campbell believed that at the confluence of the two rivers, the

Pelly was the main stream and the Yukon its tributary. He named the Yukon River upstream of Fort Selkirk the Lewes River after the chief factor of the Hudson's Bay Company, John Lee Lewes.

Bell on the other hand recognized that he had entered a much larger stream than the Porcupine. He named his discovery the Youcon. Like the Russian Glazunov, Bell used the name as it was his interpretation of the name the Natives used. Again, Bell's name also translated from the local dialect as "Great River." This naming of the river was perhaps Bell's greatest contribution to northern history as this name, of course, stuck not only as a name for the river but also for the territory as a whole.

In 1844, the headwaters of the Yukon River south of the Pelly (upstream) remained relatively unknown. The Chilkat Indians on the Alaska coast held an exclusive, self-endowed right of entry into this part of the country—a right they had no qualms about enforcing. They denied anyone access via the coastal passes and furthermore all perpetrators were forcefully removed as was later evidenced by the destruction of Campbell's Fort Selkirk at the mouth of the Pelly River.

Inevitably, of course, the Chilkats were unable to stave off progress and in 1880 the first-known group of miners entered the headwaters of the Yukon River via the Chilkoot Pass. Passage through was negotiated with the Natives on the understanding that none in the group would interfere in the Chilkats' trade.

Over the next two years, miners and prospectors compiled considerable knowledge of the Yukon's headwaters and its major tributaries such as the Teslin, Big Salmon and Stewart rivers.

In 1883, U. S. Army Lt. Frederick Schwatka floated the river from beginning to end. In his travels, Schwatka went on a naming spree. His consequent report included not only the first maps and topographical information on the river, but also all of the names he had chosen to bestow on the landmarks along the way.

Canadian geographical parties surveying the river during the next decade disallowed a number of these names in favor of those already in local use. However, many of Schwatka's names were allowed to stand and it is generally acknowledged that he is responsible for the first total documentation of the entire Yukon River route.

As reports of potential mineral riches spread "outside," more and more miners headed north and more of them started to winter in the territory. Trading posts and winter quarters were built and a general atmosphere of anticipation gripped the land. Side streams were explored, sandbars panned and the first river steamer appeared on the Lower Yukon River. The settlement of Forty Mile was established below present-day Dawson City. People arrived and stayed in such numbers that law and order was deemed necessary. The Northwest Mounted Police were sent north. Their posts were established at most of the major tributaries of the Yukon River.

In 1896, George Washington Carmack with friends Tagish Charlie and "Skookum" Jim Mason made the discovery that changed the Yukon River Basin for all time. The Klondike gold rush had begun! Suddenly more than 50,000 people rushed north and descended on the river valley in search of riches. And in the year 1898, a motley armada of more than 7,000 boats in all sizes and shapes descended the Yukon River en route to the Klondike.

Traveling upstream out of Five Finger Rapids on the *Anna Maria.*   Photo: Harry Kern.

The Yukon River became the main highway into central Yukon, and over the next fifty years some 250 stern-wheeler steamboats plied its waters. Villages and towns sprang up along its banks and acres of forest were cut to satisfy the need for fuel and building materials.

In the 1950s, a series of all-weather highways was completed throughout the territory. The riverboats were mothballed in favor of the more reliable, and year-round, road transport. River homes and whole villages were abandoned with their occupants moving to more accessible highway locations. The river no doubt breathed a sigh of relief as it was slowly allowed to return to its pre-gold rush, natural state.

Today, only scattered signs of this past commerce remain in evidence and the recreational traveler may find it hard to visualize the river as once one of the most important trade routes in the world.

The Upper Yukon River has nearly rallied to its pre-1880 state. Here and there, shelter is provided by a cabin or like-structure built by past river dwellers, but for the most part the traveler can still experience the same thrill of discovery that the early explorers felt when first embarking on this remarkable waterway.

· · · · · · · · · · · · · · · · **2** · · · · · · · · · · · · · · · ·

Traveling Tips

Remoteness

Once you leave the village of Carmacks, be prepared to handle your own emergencies.

On this leg of the Yukon River you do have limited road access at such places as Tatchun and Minto. In the event of an emergency you might also be able to obtain some assistance at Fort Selkirk and at some of the creeks closer to Dawson City, but outside help should not be counted on in the planning of your trip.

Fort Selkirk may have radio telephone communication available for most of the summer—but do not count on it. Coffee Creek, Ballarat Creek and Stewart Island are generally occupied during the summer months. In recent years, there have been several supply barges operating on the river between Minto Landing and Dawson City. These

Be prepared to handle your own emergencies.   *Photo: Gus Karpes.*

barges run infrequently and supply the placer mining operations along the river. They have no apparent schedule, but in all likelihood you will encounter at least one of them and smaller motorized river traffic as you get closer to Dawson City.

I reiterate that you should be able to handle your own emergencies. If for some reason you are unable to carry on, under no circumstances consider trying to walk out. Stay put and wait for other river traffic.

Carry one or several disposable lighters on your person at all times. The simplest method of packing a few emergency supplies is in a "fanny pack" that is worn around your waist. Irrespective of what else you choose to pack, it should contain mosquito repellent and a disposable lighter.

Garbage

If you can carry it in, you can carry it out! Buy some heavy duty garbage bags as part of your supplies. Burn leftover food. Scorch cans, flatten them and carry them along. They're a lot lighter than they were when full and won't take up any more room than they did when you first packed them. Do not bury your garbage and do not leave it in the fire pit when you take off. This is not an acceptable practice! Contrary to popular opinion, aluminum foil does not burn nor does it deteriorate. Clean up your camp fire and remove these items prior to leaving the site. Once again, if you carry it in, take it out and try to take any extra that may be around. If we all do this a time or two, the countryside can only look better for generations to come.

Camping

The Yukon River, and for that matter most rivers in the territory, is one of the last recreational rivers in North America where you can still camp where you wish. Sometimes the natural setting of a site or its distance from a given point, dictate a common stopping place and if you stop at Tatchun, Minto and Fort Selkirk you are required to camp in a specific place set aside for river travelers. Enjoy this privilege—do not abuse it. Do not indiscriminately start hacking away at the bush. Trees, even willows, take a long time to grow in Yukon's harsh climate. Using fresh evergreen bows for a bed is not nearly as comfortable as we were led to believe by our forefathers. Cutting fresh trees for a bed is unacceptable behavior. It may kill the tree and also leave a mess for the next person to clean up.

Do not take over a cabin without an invitation from its owners! Treat all property at a homestead site as private property, no matter how menial or insignificant it may appear.

Clean up your camp site before you leave. Use a small trowel or folding shovel to cover excrement and burn all toilet tissue. Women please do not bury your sanitary napkins at any time. Please note my comments under the heading Bears.

Having a camp fire is another privilege that should not be abused. If invited to share a camp, share the fire as well. Do not create another fire pit and start another fire.

Arctic grayling for breakfast.

Photo: Gus Karpes.

Hunting and Fishing

There is nothing tastier than a freshly caught Arctic grayling. It is an experience that everyone should have at least once in a lifetime. Although this stretch of the river is not renowned for its abundance of any fish, the Arctic grayling, the northern pike and some species of whitefish can be caught. Try the clear side streams for pan-sized Arctic grayling and the shallower water and sloughs around the islands for the northern pike. A fishing license is required and there are catch limits. Catch only what you intend to eat and do not leave fish offal and remains laying about the camp to attract bears and other scavengers.

I understand that during the last few years there have been several visitors writing about and filming the "living off the land" Yukon lifestyle which encourages simply moving in, building a log abode and hunting moose, bear and other species of wildlife at will. This is totally inappropriate! Hunting for any species of wildlife during a summer canoe trip is unacceptable. Anyone other than a Yukon resident must employ a registered guide, even during hunting season; and then there are regulations as to area, species and numbers, etc. The indiscriminate killing of wildlife and the cutting down of trees is not only unwelcome behavior, but also against the law of the Yukon. Anyone caught doing these activities is subject to severe fines, imprisonment or worse.

At the time of this writing, it is legal for anyone to carry a "long" gun for protection while traveling the river. This means any legal firearm, other than a handgun, is acceptable. However, this does not mean that it gives one hunting privileges. The laws governing the ownership and use of firearms in Canada and the Yukon are under constant review and you should make enquiries prior to setting out.

Drinking Water

Recently the affliction known as giardiasis and commonly referred to as "beaver fever" has become a concern. *Giardia lamblia*, the cause of it all, is found worldwide and is one of the most commonly reported human intestinal parasites. Although it can be transmitted on food and from person to person, it is frequently spread through surface water. This parasite is carried by all mammals, but the beaver seems particularly susceptible to it, hence the name beaver fever.

The symptoms of giardiasis are abdominal bloating, cramps, diarrhea and a general feeling of discomfort. The incubation period after ingestion is from ten to fourteen days. These symptoms, therefore, may not appear until your trip is over and you have returned home. The best way to avoid it is to treat your drinking water. Use tablets sold for this purpose and follow the directions or use tincture of iodine sold at most drug stores. Directions for the use of the latter are normally written on the bottle. Consider carrying one of the many water filters now on the market.

Keep clean! There is no reason to drop your normal sanitary habits during a wilderness trip. Biodegradable, liquid soap is sold in handy plastic tubes which slip easily into a shirt pocket. Use it before handling food.

Water Levels

Naturally, the level and volume of water affect the speed of the current and hence your forward progress.

Keeping in mind that you are passing a number of different major rivers on your way to Dawson City, you will realize that it is almost impossible to forecast the exact timing of the ice breaking, the river freezing, and the highs and lows during a summer season. If the coastal mountains have received more than their normal snowfall, there could be considerable water in the Yukon River. If the mountain ranges to the east receive a cold and late spring, the Pelly and Stewart are affected. If departing from Carmacks, check with the locals. They will be able to tell you where the water level is at.

Bears

I have included this specific heading in every book as it is by far the most common subject for discussion by all who travel the river. In the more than thirty years that I have spent on the rivers, I have never had a problem or a serious incident involving bears. Of course, this does not mean that bears don't exist and that I have not seen them during my travels, and it does not mean that incidents have not occurred. Bears have lived in the Yukon for thousands of years. We can avoid or at least minimize the chances of an encounter and/or an injury by understanding some of their characteristics and their lifestyle.

The most common sighting or encounter is with the black bear (*Ursus americanis*). There are an estimated 10,000 of these animals throughout the territory. They are concentrated mostly in the central and southern Yukon. Although we refer to this species as black, it can range in color from pure black through various shades of brown to cinnamon or blonde. The coloring is more consistent on the black bear than the multicolors common to the grizzly bear. The male black will reach an adult weight of about 110 kg (about 250 lbs.), while the female at maturity will weigh about 75 kg (170 lbs.).

The grizzly bear (*Ursus arctos*) ranges in color from dark, rich brown to an almost blonde. The coloring is not consistent, with the legs generally being darker. The longer body hairs particularly around the mane, have lighter tips, hence the name "silvertip" or "grizzly" for its consequent grizzly appearance. There are an estimated 6,000 to 7,000 grizzlies spread throughout the entire Yukon Territory. It is a considerably larger bear than the black. The average adult male can reach about 200 kg (450 lbs.). A female grizzly is about half that size.

Bears will eat almost anything. Although a typical diet depends on its territory, it can consist of roots, berries, grubs, small rodents, eggs, carrion and, of course, fish.

Generally speaking, all bears tend to range at the higher elevations. This is particularly true for the grizzly which prefers the subalpine areas. The river trails are visited during the early spring and during the salmon spawning season later on in the summer. As the snow line recedes, the bears tend to range higher in the more prolific growth areas where heavy bush does not hinder the growth of their primary food sources.

Bears will always investigate new food sources and have an acute sense of smell and hearing. Contrary to popular belief, a bear has good eyesight but tends to identify by its better senses of smell and hearing, thereby creating the impression that its sight is poor.

A "spoiled" bear is one that has learned to identify humans as a food source. Generally speaking, bears sighted or encountered during a wilderness river trip have not made this association and are considered wilderness bears.

Following are some suggestions for campers traveling through bear country.

1. When carrying fresh foods such as fruits, meats, vegetables or sweets, carry them in an airtight container.

2. Cook enough food for one meal only, and if there are leftovers do not leave them out. If you cannot put them in an airtight container, burn them immediately.

3. Scorch all noncombustible materials such as cans and tinfoil before putting them in your garbage container.

4. Meticulously clean up after a meal. Use a little bleach or chlorox in your dishwater.

5. Do not store foods in your tent.

6. Clean fish away from camp if possible. If cleaning them in camp, make sure that all traces of the innards and leftovers are cleaned up and incinerated.

7. Never bury garbage, pack out everything that you brought in. If you were able to pack it in, there should be more than enough room to carry it out!

8. Personal cleanliness is also important. A pair of pants you've worn and wiped your hands and knife on starts to smell quite good to a bear on day ten. Bring a plastic, airtight laundry bag for soiled clothing.

A word of caution to women. Try to stay out of bear country during your menstrual cycle. There is some evidence that bear is attracted to women during this physiological condition. If this is unavoidable, keep scrupulously clean. Acquaint your partner or fellow campers with your problem so that you will have the time and privacy to deal with it. Incinerate used sanitary napkins or tampons as soon as possible. Carry a supply of ziplock plastic baggies and some paper lunch bags with you. These will do as temporary receptacles en route until a fire is available.

Pay attention to your surroundings. Patrol a campsite prior to setting up camp. If the spot is an obvious feeding area for a bear, continue on to the next suitable site. You can smell a bear in the woods. Their excrement and body odor is very pungent, particularly in late summer.

Bear Encounters

Bears, like dogs, all have their own personality, and there is no single answer to the handling of a bear encounter. Generally speaking, a bear will avoid an encounter with a human if it knows you are there and you are not competing with it for food. Let the bear know you are there.

A bear standing up, slapping its sides and waving its nose around is trying to identify you. Keep facing it and do not turn your back and run. Talk to it and announce your presence, slowly retreating and giving it room. Do not "trap" a bear, and by this I mean, always give it an avenue of escape. Do not contest the right to food even if you are the one that bought it at the supermarket. Never attempt to get close to a bear cub. Although mama may not be visible, she is around and will do anything to prevent harm coming to her cubs and you present a threat to them.

I have found that all the other animals in the bush can warn us in advance. Chattering squirrels, ravens, jays, etc., can herald the approach of a bear or other large mammal such as a moose. Prior to the large animal entering camp, all the normal noises seem to reach a crescendo, followed by an almost eerie period of total silence before the large animal announces itself.

To conclude, I do not wish to create the impression that bear problems are inherent to and an everyday hazard of a wilderness canoe trip. Bear problems are rare. Sightings are not a daily occurrence. With these comments, I hope some of the mystery has been dispelled. Bears are quite willing to share the wilderness with you as long as you recognize their right to be there. Sightings of bears and other wild animals are something to be enjoyed as a rare bonus to the wilderness experience and with proper precautions should not create a problem.

The Major River Sections

Carmacks to Rink Rapids—30 miles (48 km)

MAPS 1, 2	PAGES 28, 29

Upon leaving Carmacks the river twists and turns every which way and it will appear as if you are not making any progress whatsoever. The same Carmacks hills keep appearing on the skyline. Collectively, the first few curves out of Carmacks are known as Shirtwaste Bend because of the way they resemble the split tails of an old shirt.

During the annual run of salmon, which occurs in late July and August, there are always a number of fish camps in the first ten miles (16 km). Most of them are of a temporary nature and only occupied during the fishing season. I have only marked the most prominent ones. Although the Nordenskiold does not add all that much water, the river does appear to gain in stature almost immediately upon leaving Carmacks. High clay banks and rocky bluffs now start to appear on both sides of the river which scenery does not change appreciably until you are about to go through Five Finger Rapids below.

Five Finger Rapids, twenty-four miles (38 km) downstream, is probably one of the most famous and most photographed spots on the Yukon River. Our first description of the rapids comes from Lt. Frederick Schwatka in 1883. The Indians traveling with him warned him of the perils ahead which prompted him to scout ahead on foot. "I found them to be a contraction of the river bed, into about one-third its usual width of from four to six hundred yards, and that the stream was also impeded by a number of massive trap rocks, thirty to forty feet high, lying directly in the channel and dividing it into three or four well marked channels, the second from the east, being the one ordinarily used by the Indians."

I don't know what "trap" rocks are, but I assume he meant that they formed a natural trap or danger to river traffic.

Simply put, the river is blocked by a number of large rocks which divide the river into specific channels or "fingers." There are five openings, hence the name. Schwatka rejected the channel supposedly used by the local Indians in favor of the most eastern (right) channel. He continues: "We essayed the extreme right-hand (eastern) passage, although it was quite narrow and its boiling current was covered with waves running two or three feet high."

You should also be on the right side of the river as you approach Five Finger Rapids. The channel you want is the narrow channel against the right (east) shore. Stay in the middle of this channel where the water coming off both rock walls forms a "chute." There are some standing waves, but the passage through is short. Keep the canoe moving. Once through, follow the volume of water to the left. Pass to the left of the island directly below the rapids. Follow the side of the island and, once past, swing to the right into Tatchun Creek. I repeat the warning: Be on the right side of the river as you approach Five Finger Rapids!

Rink Rapids, named for Dr. Henry Rink of Denmark, is six miles below Five Finger Rapids. It is not nearly as demanding to navigate unless you are in a powerboat. In that case, I always find that I have more problems with Rink because of the shallow water and the even narrower channel. Canoes and small powerboats should again approach the rapids on the right side of the river. You will have lots of time to line up. As you approach the rapids, it will appear as if the right side of the river is full of white and turbulent water, but the narrow channel up against the right shore becomes clear as you get closer. For powerboats the next mile is very shallow and has a number of rocks in it. Take your time.

Finding a camp is difficult in this stretch of river as the shoreline is predominantly very steep. There are some possible camps on the islands just past Murray Creek and there is some level ground at the old Five Finger Coal Mine. If it is late in the day when you're going through Five Finger Rapids, stop at the Tatchun Creek campground as you will not have another opportunity for a camp until you are through Rink Rapids, six miles further on. If you have the time, I recommend that you proceed through both Five Finger Rapids and Rink Rapids before stopping for the night. Immediately below Rink Rapids there are a number of islands that can suffice as an overnight camp. There is also plenty of room at Yukon Crossing, seven miles below Rink

Rapids. Unfortunately, you will be subjected to road noise as the Klondike Highway runs close to the river between Rink Rapids and Yukon Crossing.

Three miles below Rink Rapids, against the right sandy bluff you will see a large pocket of volcanic ash commonly referred to as Sam McGee's Ashes. This and the fine layer of ash prominent at the top of most cut banks occur throughout the Yukon as a result of volcanic eruptions in the St. Elias Range of mountains to the west. Some very violent eruptions occurred around 900 to 100 A.D. and as recent as the eruption of Mount Katmai in 1912.

Rink Rapids to Fort Selkirk—52 miles (83 km)

MAPS 2–6   **PAGES 29–33**

The character of the river again undergoes a change. This section is distinctive for the numerous islands. The vegetation also seems to be more prolific than what has been encountered so far and this trend will continue all the way to Dawson City. The configurations of the islands on the map are difficult to physically relate to what you see. Gravel bars and sometimes whole islands can appear and disappear during a spring breakup. Keep your eye on the shoreline. The leaning and skinned trees and bushes are evidence of the havoc that ice can cause.

Choosing a camp in the evening will be much easier for the remainder of your trip. The islands will stay with you until you reach Dawson City and their number will give you endless choices for a campsite. Driftwood for a fire is never far away.

This stretch of river was more demanding for the river pilots of old than any other section of the river between Whitehorse and Dawson City. The first problems encountered were in the Minto Flats. Here, the channel was constantly on the move, never the same from year to year.

Next came the run into Fort Selkirk with such encouraging names as Devil's Crossing, Hell's Gate and Slaughterhouse Slough. This was an extremely trying stretch of river and it was common to see several of the large boats stuck in the "Gate." Public Works Canada and the British Yukon Navigation Company spent a lot of time and money in maintaining a channel through here. They blasted rocks out of the channel, built several wing dams to control the water flow and even drove ring bolts into the rock wall. A block and cable could be attached to the large ring bolts and the ship's winch deployed to pull the vessel off the gravel bars or to assist the vessel in traveling against the fast current running through the gate. Watch the rock wall as you come through. Some of the anchors are still there.

Since the disappearance of the large boats, the channel has deteriorated very badly. For powerboats it is very precarious from Minto to Fort Selkirk. It is shallow and rocky around Devil's Crossing and Hell's Gate, with the channel into Fort Selkirk presenting a real challenge. Extreme caution should be exercised at all water levels.

Up until recently the channel into Fort Selkirk was through Slaughterhouse Slough and the mouth of the Pelly River. In 1997, when I was piloting the large riverboat *Anna Maria* from Dawson to Whitehorse, I used the old steamer channel. It was very shallow but navigable, and from all reports it was better than the Slaughterhouse Slough channel previously in use. I have marked the channel into Fort Selkirk. Keep in mind that this can change very quickly.

By canoe, I would suggest that you head for the left side of the river near mile 260 (km 416), Von Wilczek Creek. You can stay on that side of the river until you get to Fort Selkirk. The creek is just past Minto Landing which is an unmistakable landmark. There is a large cleared campground and the campground shelters are clearly visible from the river. If you are carrying a bag of garbage, take time to stop and dispose of it in one the campground containers. The water in front of the landing is fast. Be cautious in your approach and landing.

Periodically there are several large barges operating in this stretch of river. Please keep in mind that these barges and, for that matter, all powerboats, are very restricted in their choice of a channel. Please yield.

Just past Minto Landing and again as you approach the Pelly River, keep your eyes on the rocky hills on the right side of the river. The area has several resident bands of white Dall sheep. Look for them against the rocky hills. Because of their bright white color, they show up quite clearly.

During the last four years, forest fires have burned acres of forest along the river. Finding a place to camp away from this damage may be difficult, particularly at the higher water levels. Try the islands just past Big Creek or the Ingersoll Islands just past Hell's Gate. Because of the rampant erosion of the shoreline and the islands, this area is renowned for sweepers, trees that topple into the water but remain anchored to the shore. If you get caught in one of these, you are literally swept out of your canoe which often capsizes at the same time, depending on the height of the sweeper above the waterline. When planning a landing, look at the area carefully. Land with the canoe facing upstream and insure that there are no sweepers immediately behind you in the event you drift backwards before making fast to the shore.

.

Fort Selkirk to Stewart Island—108 miles (172 km)

MAPS 6–16   **PAGES 33–43**

Historically speaking, you are now coming into what is perhaps the most interesting part of the Yukon River for its entire distance to the Bering Sea. From Fort Selkirk to Forty Mile, the pre-gold rush village below present-day Dawson City, all recent development and history involved gold mining one way or another. Robert Campbell and his followers, the Chilkat Indians and the local Natives were for the most part involved in the fur trade and there are unconfirmed tales of gold finds being deliberately ignored and suppressed for fear that they would destroy this trade and bring unwanted people and development to the country. Most, if not all, of the creeks and rivers from here to the Klondike have experienced their own mini gold rush during the last century.

As you leave Fort Selkirk, the basalt palisades so prominent opposite the village, prevail for about another twenty-five miles before a notable change of scenery takes place. As you can see on the map, there are many islands in the river from here to Dawson City. Finding a camp in the evening should not present a problem, and due to the width of the river and the number of islands, it is also quite possible to pass another boat or canoe and not even be aware of it.

It is difficult to find most of the creeks that flow into the river. They enter the Yukon through a number of backwater sloughs away from the main channel. As a consequence it is tricky to locate yourself on the map. Excelsior Creek at mile 344 (km 550) is the exception. For the others, a riverside clearing, a road leading off into the bush, a fuel tank or other such signs of civilization on the riverbank is the only way you will know you are at one of the creeks marked on the map.

There are permanent residents at most creek locations beyond Britannia Creek. It is not a good idea to impose on them or plan an overnight stay at the creeks with some notable exceptions such as

Excelsior and Britannia creeks. These locations are described in the editorial portion of the book.

Should you stop at a location that is obviously occupied, please respect all private property no matter how insignificant it may appear to be. All materials take on a special meaning when you are in the middle of nowhere. The best illustration of this is to liken it to the unique affection you now feel for the equipment with which you started this trip. The special "sticks" that keep your gear off the bottom of the canoe, the reused baggy that keeps this book dry and the two-cent piece of cord that prevents the loss of your favorite cutting tool—all are now old friends and seem irreplaceable. A camping/canoeing trip away from the corner store and the everyday conveniences gives us a new awareness of items and conditions which we normally accept as just being there and take for granted.

From Isaac Creek down, most of the larger creeks have at one time been prospected and mined since the late 1800s. Quite a few of them are presently being worked.

Beyond Britannia Creek the personality of the river again changes. High hills border both banks and although the current does not notably slow down, it appears to slither and slide past silently, without so much as a gurgle, giving you almost ominous expectations. To me, no other stretch of river accents the wilderness solitude more.

More and more silt runs into the river, and as you pass the White River it becomes very obvious how it came by its name. You can hear the silt grind away at your canoe. Drinking water should be obtained from a clear side stream if possible.

The mouth of the Stewart River is difficult to spot unless you swing to the right just past the White River instead of following what appears to be the main channel. Stewart Island is an oasis in the wilderness. Unfortunately, during the past ten years, more of the island has been disappearing into the river. Do not expect too much as very little of the island remains. The Burian family who lived there has almost all moved away. Some now live in Dawson City, and, to my knowledge, Robin Burian is the only member of the family that still resides on the island. The main house followed most of the cabins into the river in the spring of 1997. I have enjoyed the hospitality of the Burian family for almost thirty years. It is sad to see this traditional part of Yukon River history come to an end.

Stewart Island to Dawson City—70 miles (112 km)

You are in the home stretch and, once again, the character and appearance of the river changes significantly. The changing color of the water is most noticeable. If you are making coffee with the river water, it definitely adds a little body, color and flavor to it. The main river and most of the creeks are very silty and the abrasive noises of the silt-laden water running past your canoe are constant. Islands, sloughs and sandbars are still scattered at random for the entire distance and the first clear stretch of unobstructed water will not come until Dawson City is in view. There is also a significant change in the appearance of the islands. You now have to be careful in choosing a camp for the night. Most islands are choked with vegetation and most are difficult to approach due to eroding banks and an abundance of sweepers laying in wait. The best bet for a camp is still on the islands where an open site will provide some respite from the bugs, especially if there is a slight breeze blowing which tends to keep them grounded. Check out the downstream end of the islands.

The additional flow of the Stewart and White rivers picks up the speed of the current. This is particularly noticeable during the flood stages in the first month of summer. With the faster current comes widespread erosion such as evidenced at Stewart Island. This again gives rise to the perils of numerous sweepers and sluffing shorelines. I repeat the warning of being cautious in an approach to a landing. Take your time to look the situation over.

There is a wilderness camp site established at Galena or Dog Creek. This is on the left at mile 434 (km 694). More on the Ancient Voices Wilderness Camp in the historical log later in the book. I haven't attempted to identify any other camping locations in the accompanying maps.

From Stewart Island to Dawson City you are in "Klondike Country." The richest of the Klondike creeks are for the most part, encompassed in the country between Indian River at mile 434 (km 694) and the Klondike River at Dawson City.

Some canoeists take this last leg of the trip in one instalment, drifting through the night and eating on board. A special bond is formed with the river as you drift quietly through the twilight hours—listening to the wilderness sounds, the silt hissing along the length of your canoe and watching the moon rise over the Klondike, however brief its appearance. The expectancy of a new day and the dawn's arrival somehow seems to provide an appropriate setting for a landing in Dawson City.

4. Maps

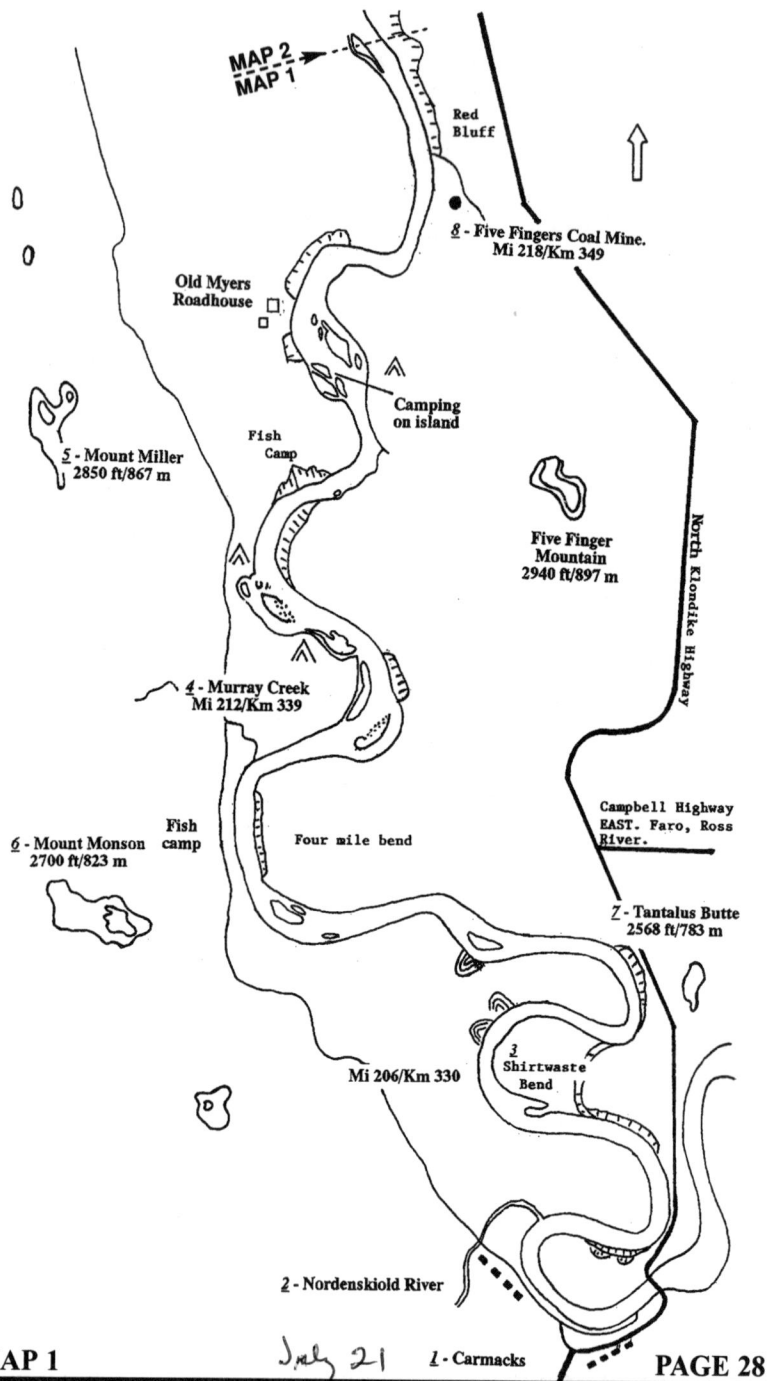

MAP 2
MAP 1

Red Bluff

8 - Five Fingers Coal Mine.
Mi 218/Km 349

Old Myers
Roadhouse

Camping
on island

Fish
Camp

5 - Mount Miller
2850 ft/867 m

Five Finger
Mountain
2940 ft/897 m

North Klondike Highway

4 - Murray Creek
Mi 212/Km 339

Campbell Highway
EAST. Faro, Ross
River.

Fish
camp

Four mile bend

6 - Mount Monson
2700 ft/823 m

7 - Tantalus Butte
2568 ft/783 m

Mi 206/Km 330

3
Shirtwaste
Bend

2 - Nordenskiold River

MAP 1

July 21

1 - Carmacks

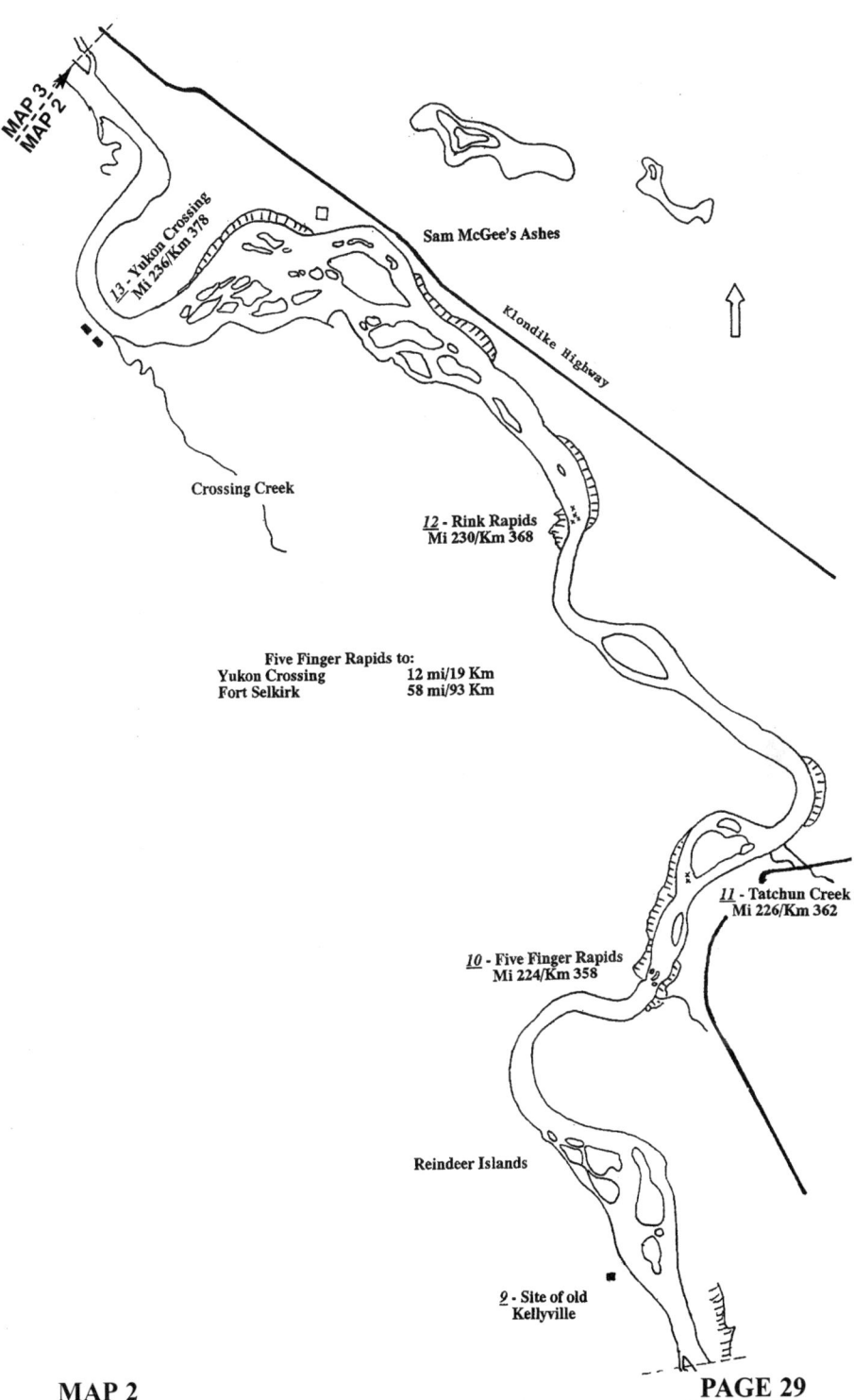

MAP 3
MAP 2

_13_ - Yukon Crossing
Mi 236/Km 378

Sam McGee's Ashes

Klondike Highway

Crossing Creek

_12_ - Rink Rapids
Mi 230/Km 368

Five Finger Rapids to:
Yukon Crossing   12 mi/19 Km
Fort Selkirk   58 mi/93 Km

_11_ - Tatchun Creek
Mi 226/Km 362

_10_ - Five Finger Rapids
Mi 224/Km 358

Reindeer Islands

_9_ - Site of old
Kellyville

MAP 2   **PAGE 29**

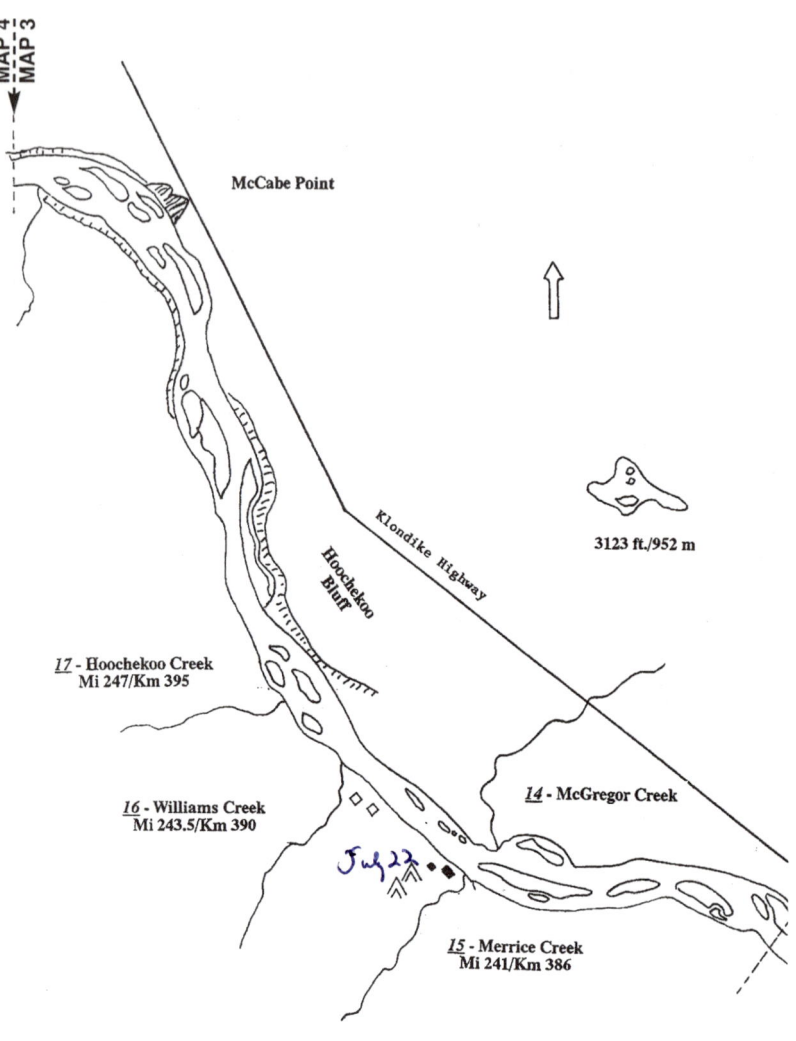

McCabe Point

Klondike Highway

3123 ft./952 m

Hoochekoo Bluff

17 - Hoochekoo Creek
Mi 247/Km 395

16 - Williams Creek
Mi 243.5/Km 390

14 - McGregor Creek

15 - Merrice Creek
Mi 241/Km 386

MAP 4
MAP 3

Williams Creek to:
Minto   15.5 mi/26 Km
Fort Selkirk   39.5 mi/63 Km
Dawson City   217.5 mi/348 Km

MAP 3   **PAGE 30**

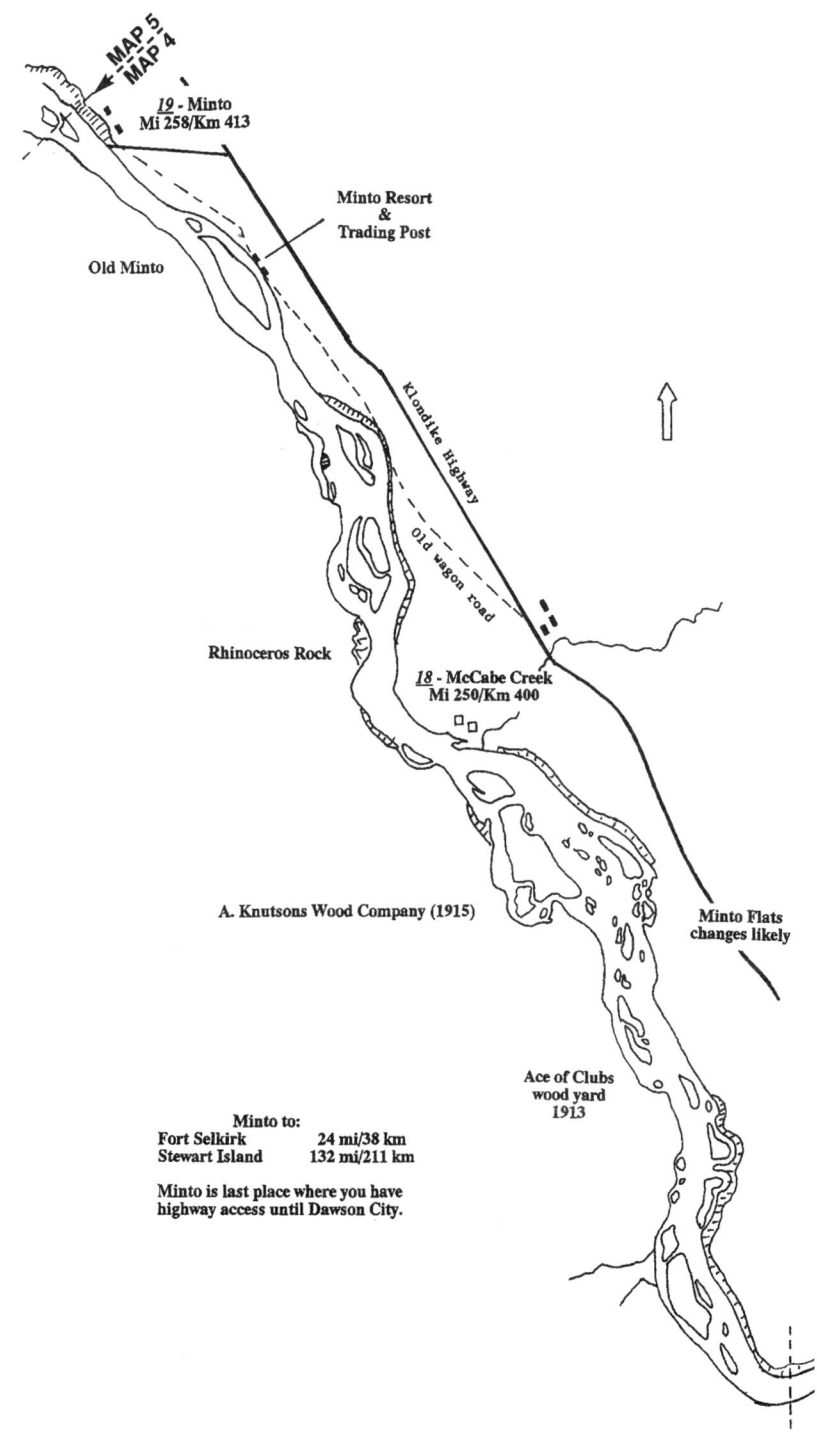

MAP 5
MAP 4

19 - Minto
Mi 258/Km 413

Minto Resort
&
Trading Post

Old Minto

Klondike Highway

Old wagon road

Rhinoceros Rock

18 - McCabe Creek
Mi 250/Km 400

A. Knutsons Wood Company (1915)

Minto Flats
changes likely

Ace of Clubs
wood yard
1913

Minto to:
Fort Selkirk   24 mi/38 km
Stewart Island   132 mi/211 km

Minto is last place where you have
highway access until Dawson City.

MAP 4

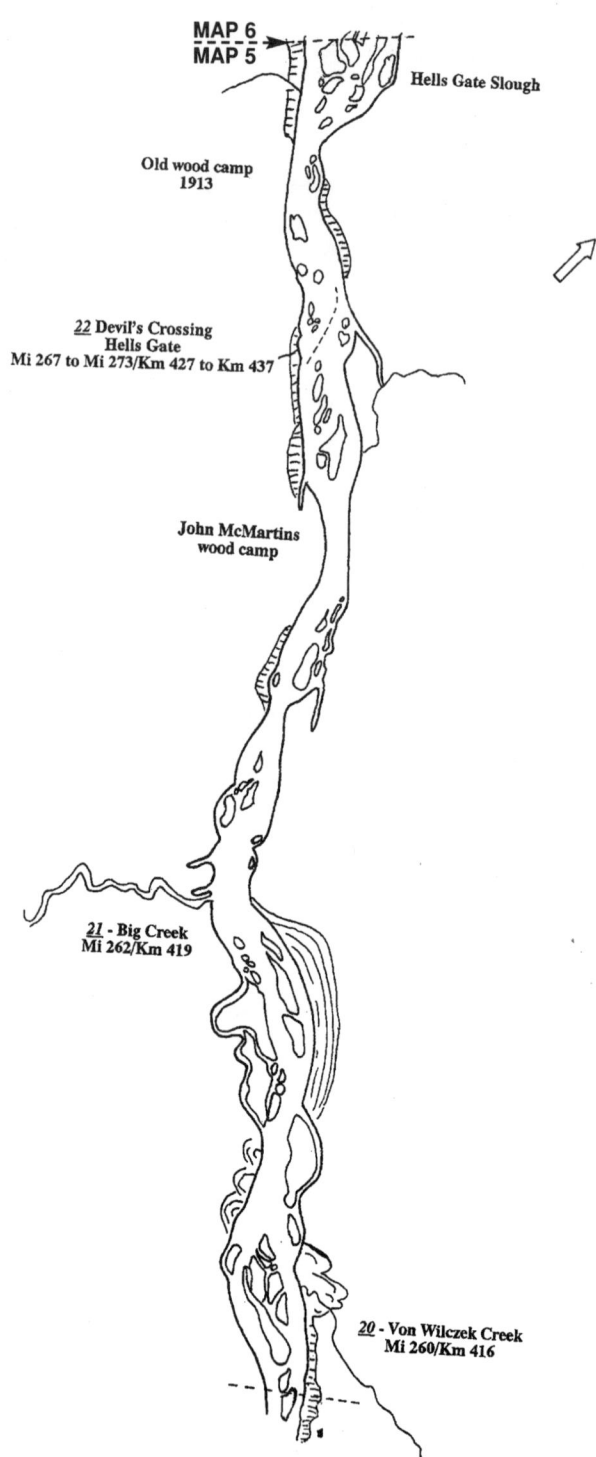

MAP 6
MAP 5

Hells Gate Slough

Old wood camp
1913

22 Devil's Crossing
Hells Gate
Mi 267 to Mi 273/Km 427 to Km 437

John McMartins
wood camp

21 - Big Creek
Mi 262/Km 419

20 - Von Wilczek Creek
Mi 260/Km 416

MAP 5   **PAGE 32**

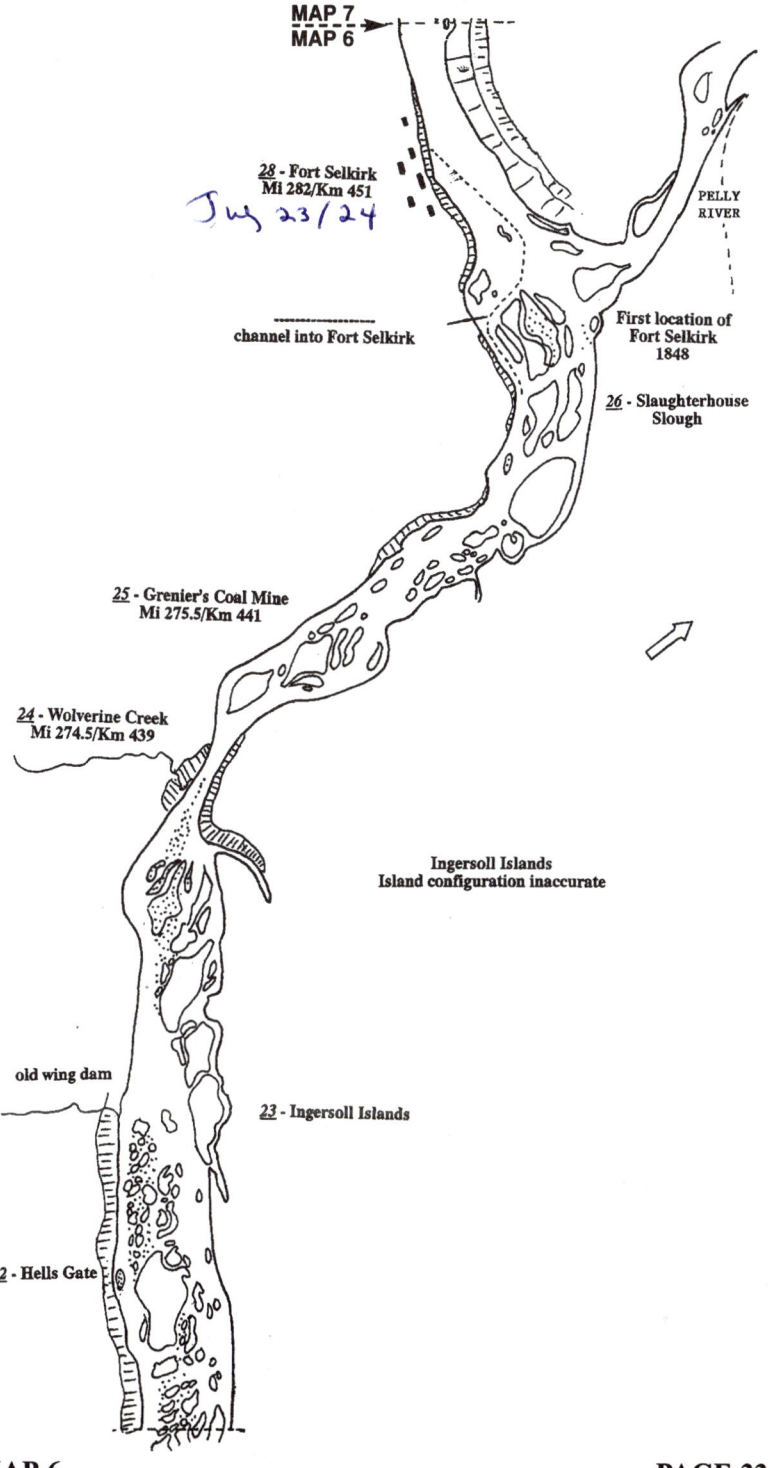

MAP 7
MAP 6

28 - Fort Selkirk
Mi 282/Km 451

July 23/24

PELLY
RIVER

channel into Fort Selkirk

First location of
Fort Selkirk
1848

26 - Slaughterhouse
Slough

25 - Grenier's Coal Mine
Mi 275.5/Km 441

24 - Wolverine Creek
Mi 274.5/Km 439

Ingersoll Islands
Island configuration inaccurate

old wing dam

23 - Ingersoll Islands

22 - Hells Gate

MAP 6

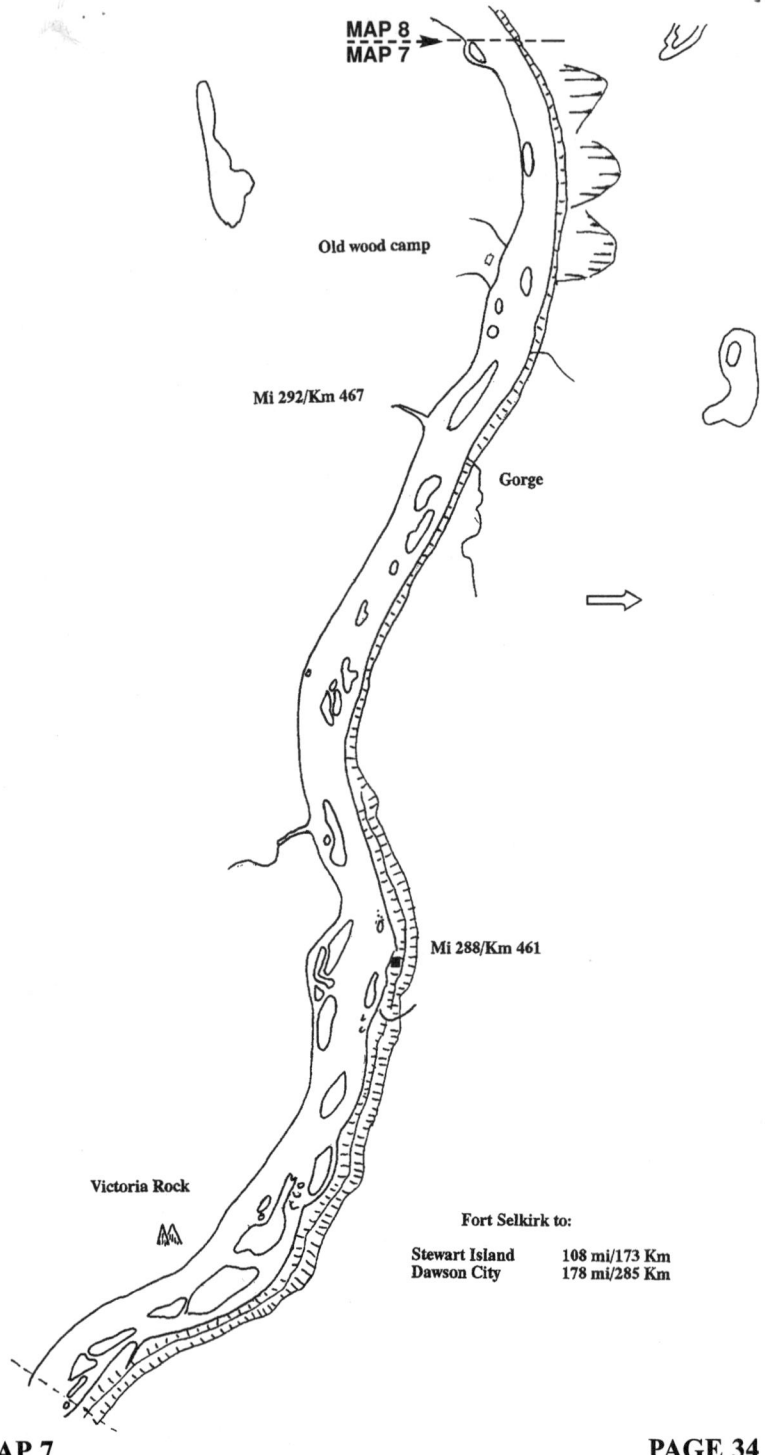

MAP 8
MAP 7

Old wood camp

Mi 292/Km 467

Gorge

Mi 288/Km 461

Victoria Rock

Fort Selkirk to:

Stewart Island   108 mi/173 Km
Dawson City   178 mi/285 Km

MAP 7

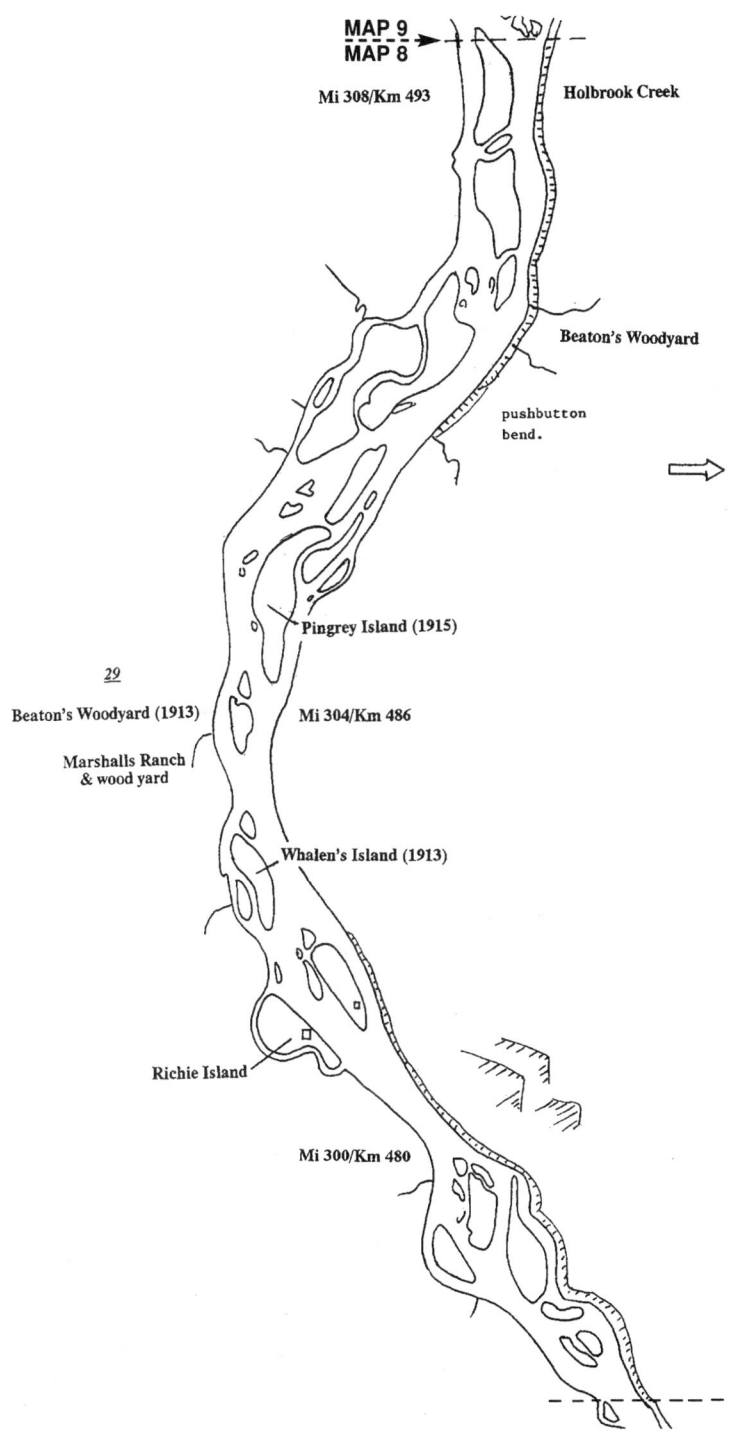

MAP 9
MAP 8

Mi 308/Km 493

Holbrook Creek

Beaton's Woodyard

pushbutton
bend.

Pingrey Island (1915)

29

Beaton's Woodyard (1913)

Mi 304/Km 486

Marshalls Ranch
& wood yard

Whalen's Island (1913)

Richie Island

Mi 300/Km 480

MAP 8   **PAGE 35**

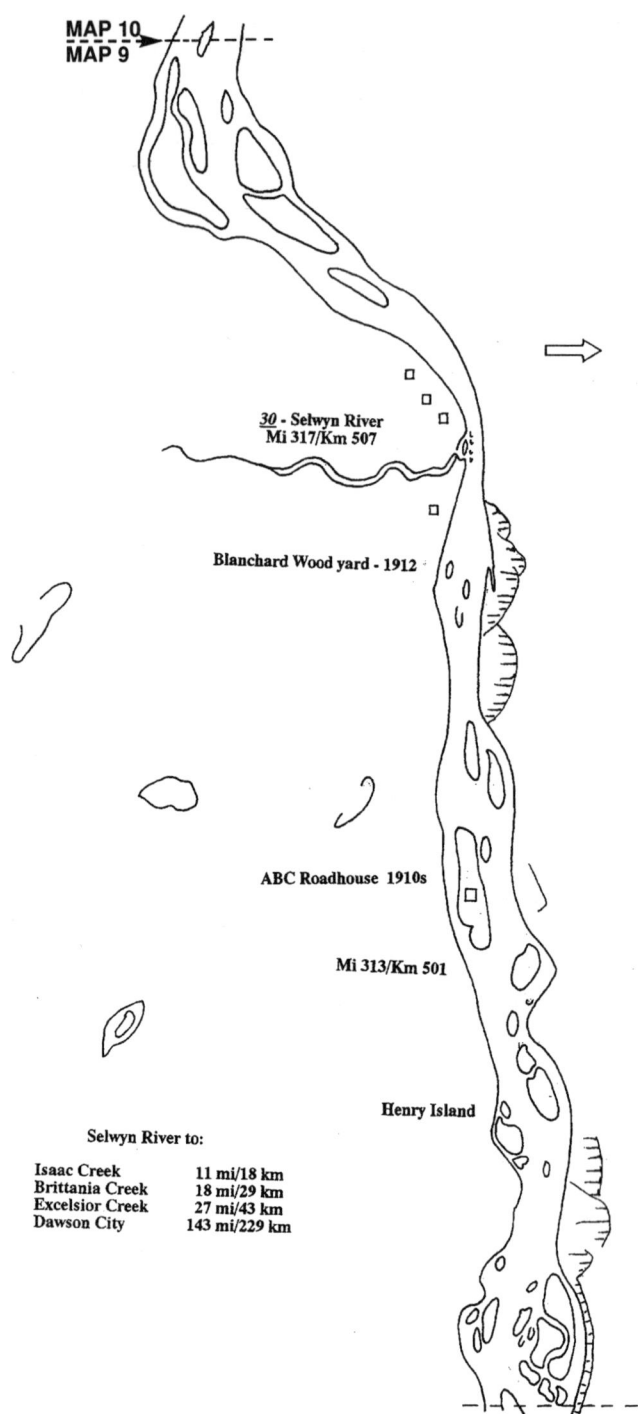

MAP 10
MAP 9

30 - Selwyn River
Mi 317/Km 507

Blanchard Wood yard - 1912

ABC Roadhouse 1910s

Mi 313/Km 501

Henry Island

Selwyn River to:

Isaac Creek	11 mi/18 km
Brittania Creek	18 mi/29 km
Excelsior Creek	27 mi/43 km
Dawson City	143 mi/229 km

MAP 9   **PAGE 36**

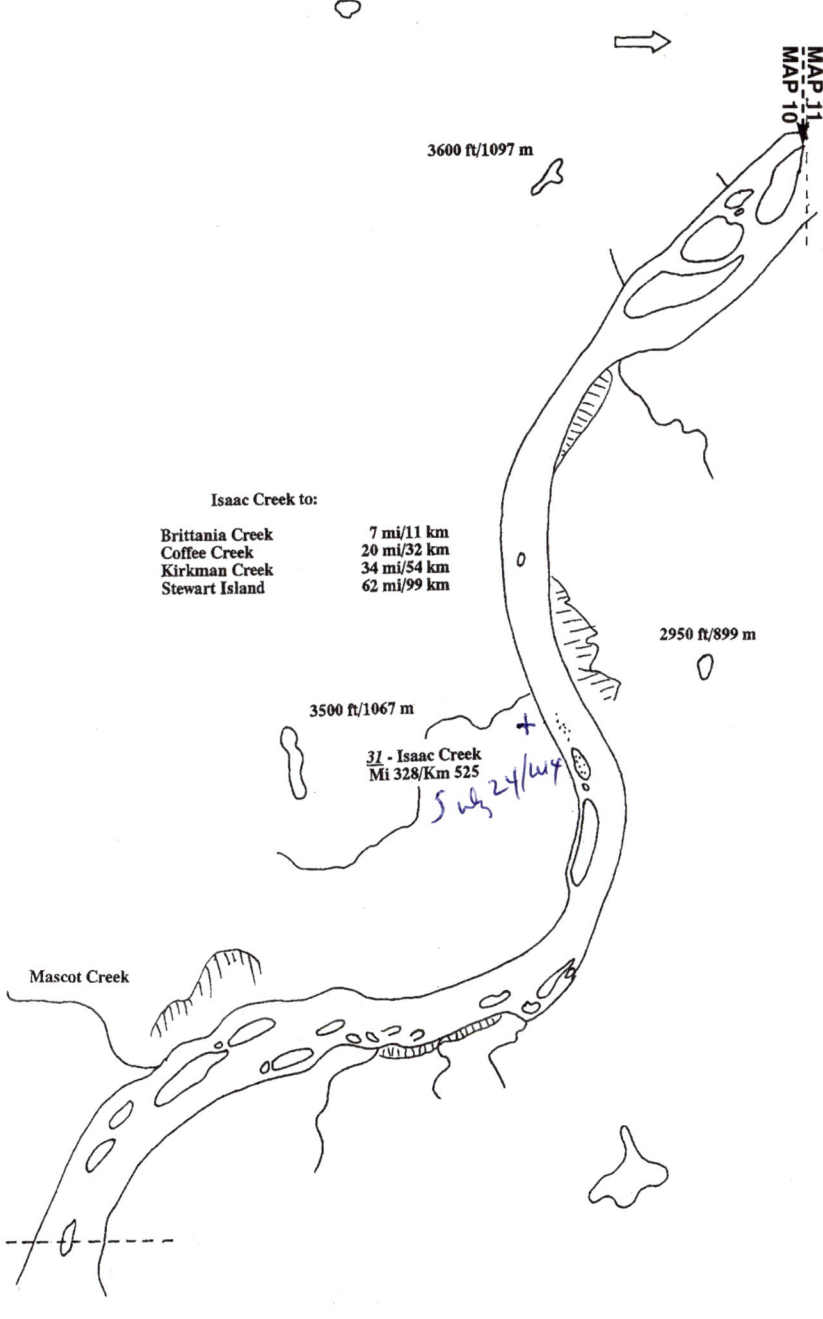

3600 ft/1097 m

MAP 11
MAP 10

Isaac Creek to:

Brittania Creek	7 mi/11 km
Coffee Creek	20 mi/32 km
Kirkman Creek	34 mi/54 km
Stewart Island	62 mi/99 km

2950 ft/899 m

3500 ft/1067 m

_31_ - Isaac Creek
Mi 328/Km 525

Mascot Creek

MAP 10   **PAGE 37**

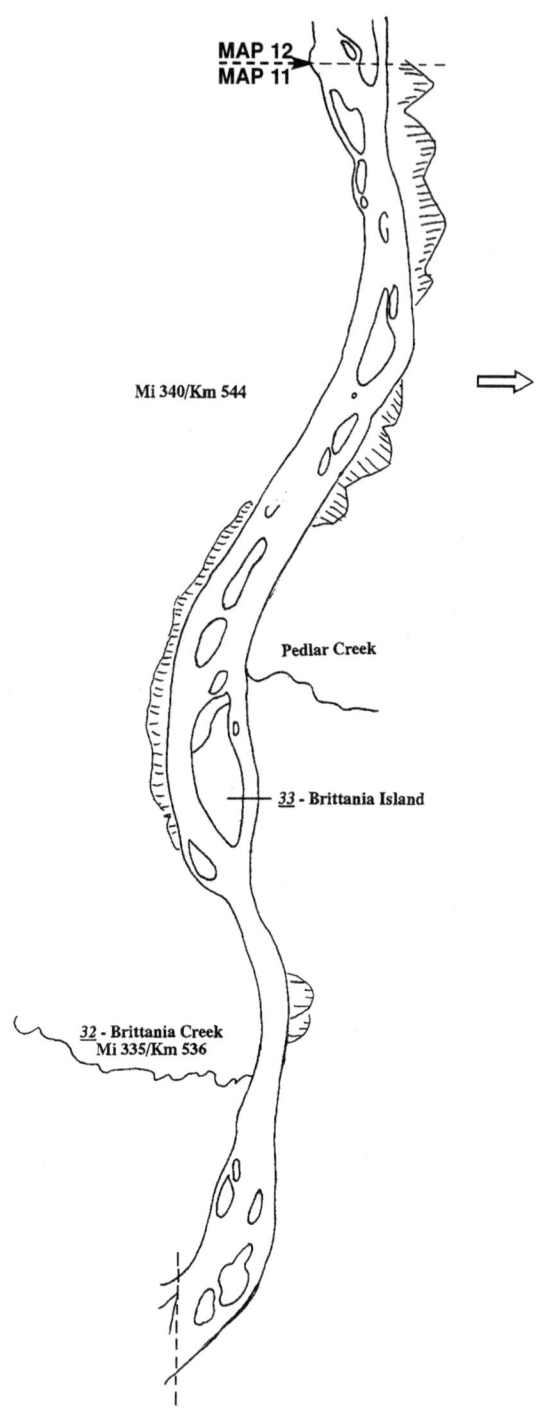

MAP 12
MAP 11

Mi 340/Km 544

Pedlar Creek

_33_ - Brittania Island

_32_ - Brittania Creek
Mi 335/Km 536

MAP 11   **PAGE 38**

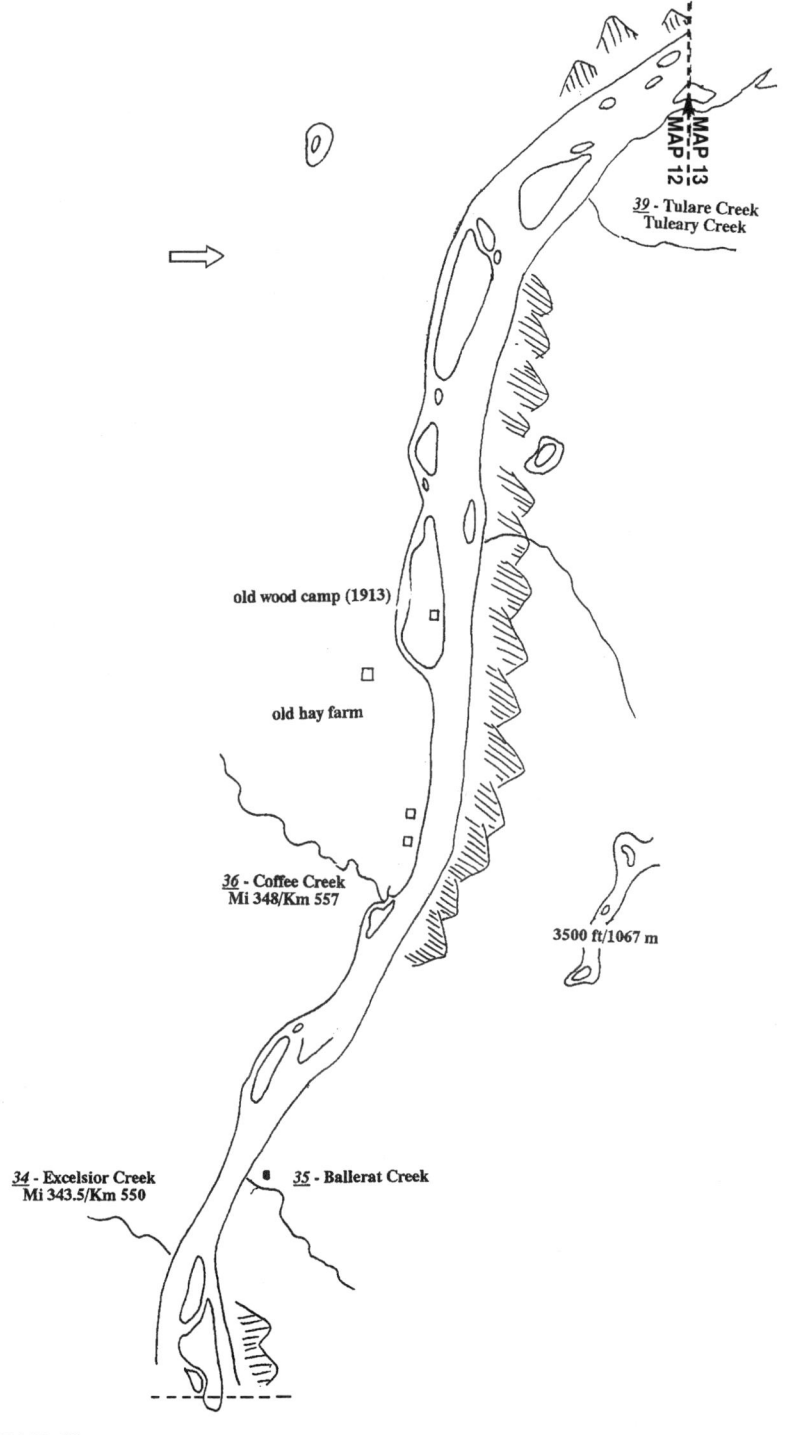

MAP 13
MAP 12

39 - Tulare Creek
Tuleary Creek

old wood camp (1913)

old hay farm

36 - Coffee Creek
Mi 348/Km 557

3500 ft/1067 m

34 - Excelsior Creek
Mi 343.5/Km 550

35 - Ballerat Creek

MAP 12

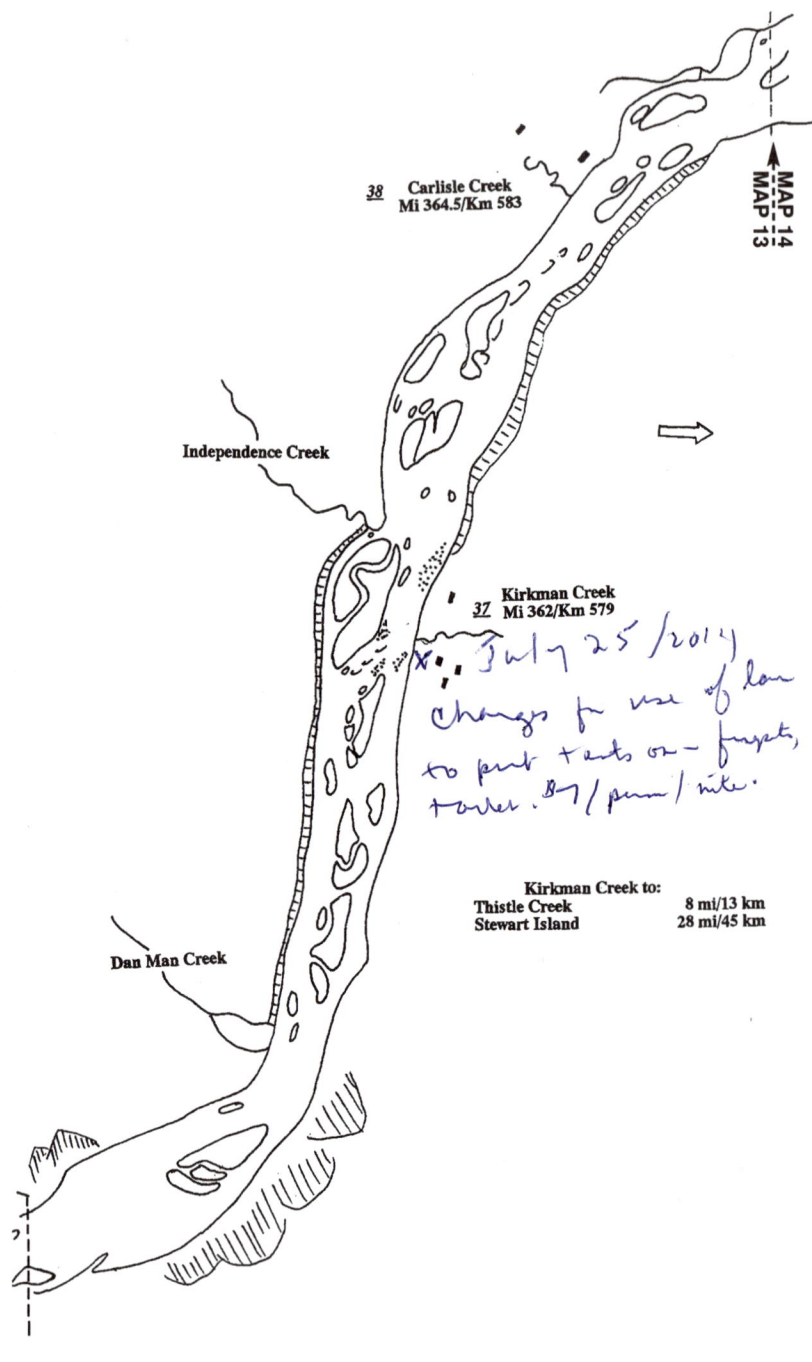

38 Carlisle Creek
Mi 364.5/Km 583

Independence Creek

37 Kirkman Creek
Mi 362/Km 579

July 25/2014
changes for use of land
to put tents on — firepits,
toilet. $7/ppsn / nite.

Kirkman Creek to:
Thistle Creek   8 mi/13 km
Stewart Island   28 mi/45 km

Dan Man Creek

MAP 14
MAP 13

MAP 13   PAGE 40

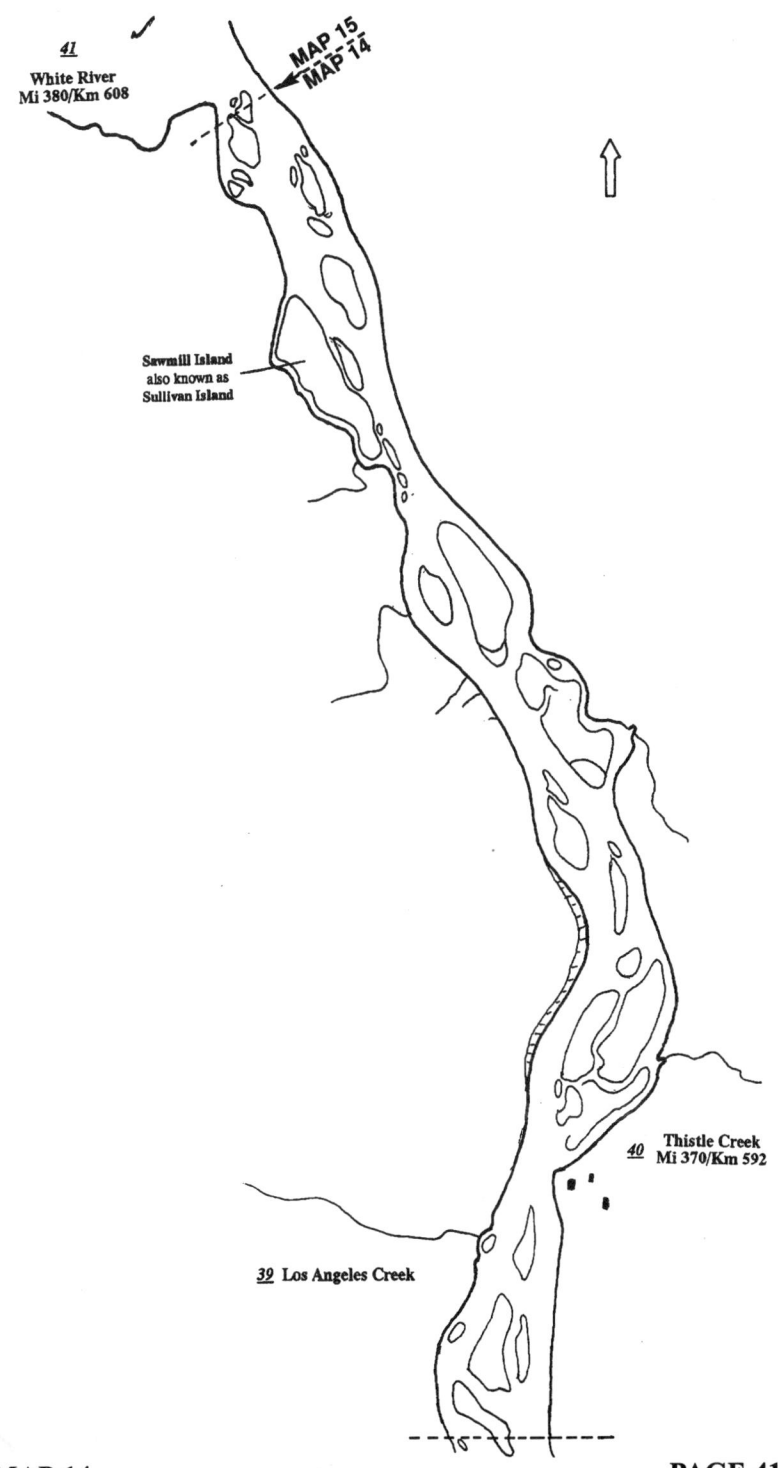

41
White River
Mi 380/Km 608

MAP 15
MAP 14

Sawmill Island
also known as
Sullivan Island

40 Thistle Creek
Mi 370/Km 592

39 Los Angeles Creek

MAP 14　　　　　　　　　　　　　　　　**PAGE 41**

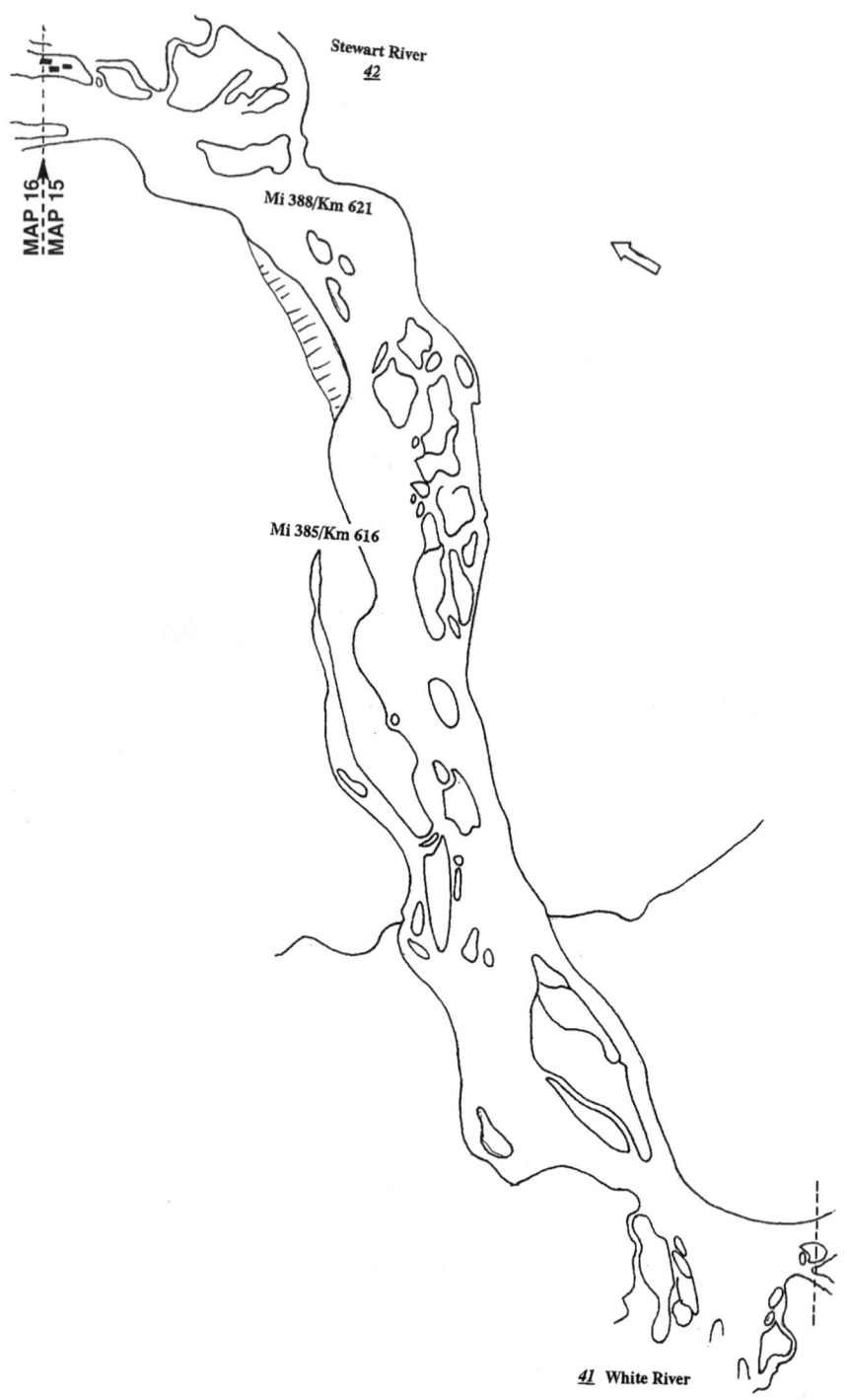

Stewart River
_42_

Mi 388/Km 621

Mi 385/Km 616

MAP 16
MAP 15

_41_ White River

MAP 15   **PAGE 42**

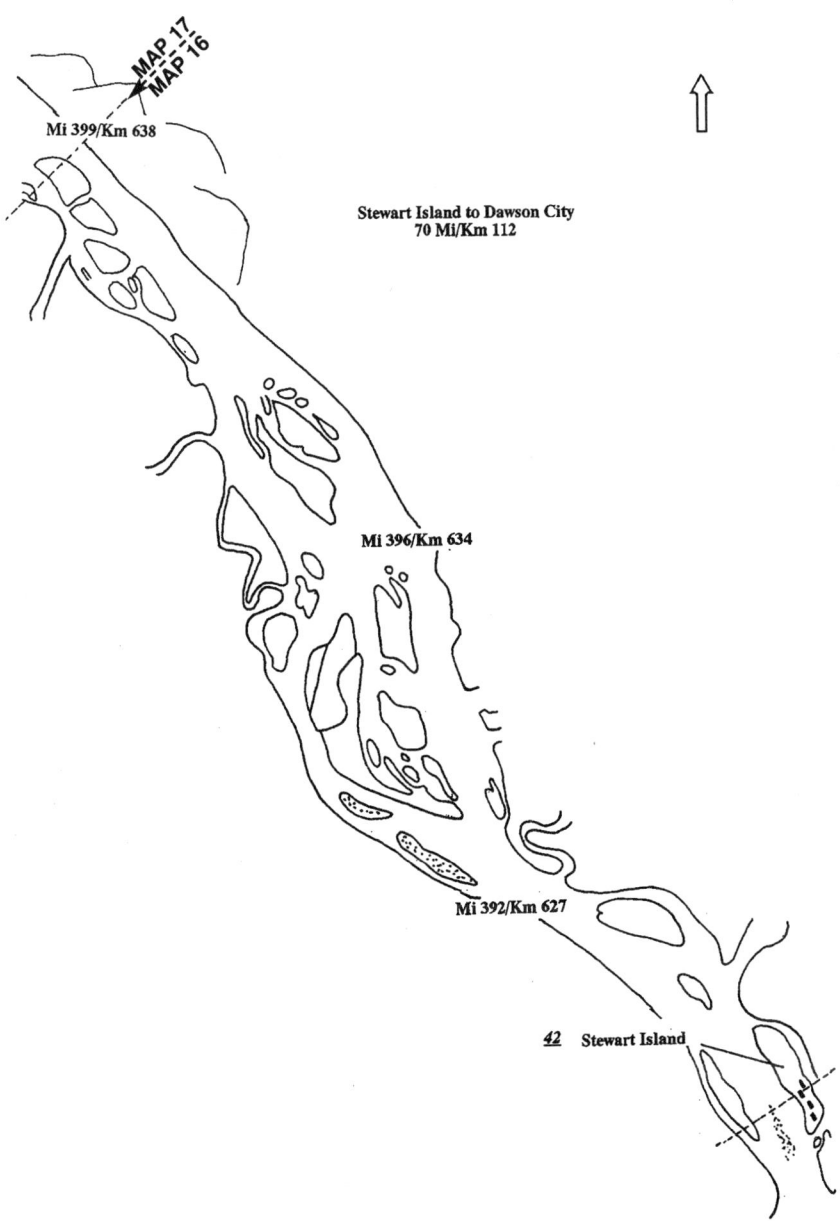

MAP 17
MAP 16

Mi 399/Km 638

Stewart Island to Dawson City
70 Mi/Km 112

Mi 396/Km 634

Mi 392/Km 627

_42_ Stewart Island

MAP 16   **PAGE 43**

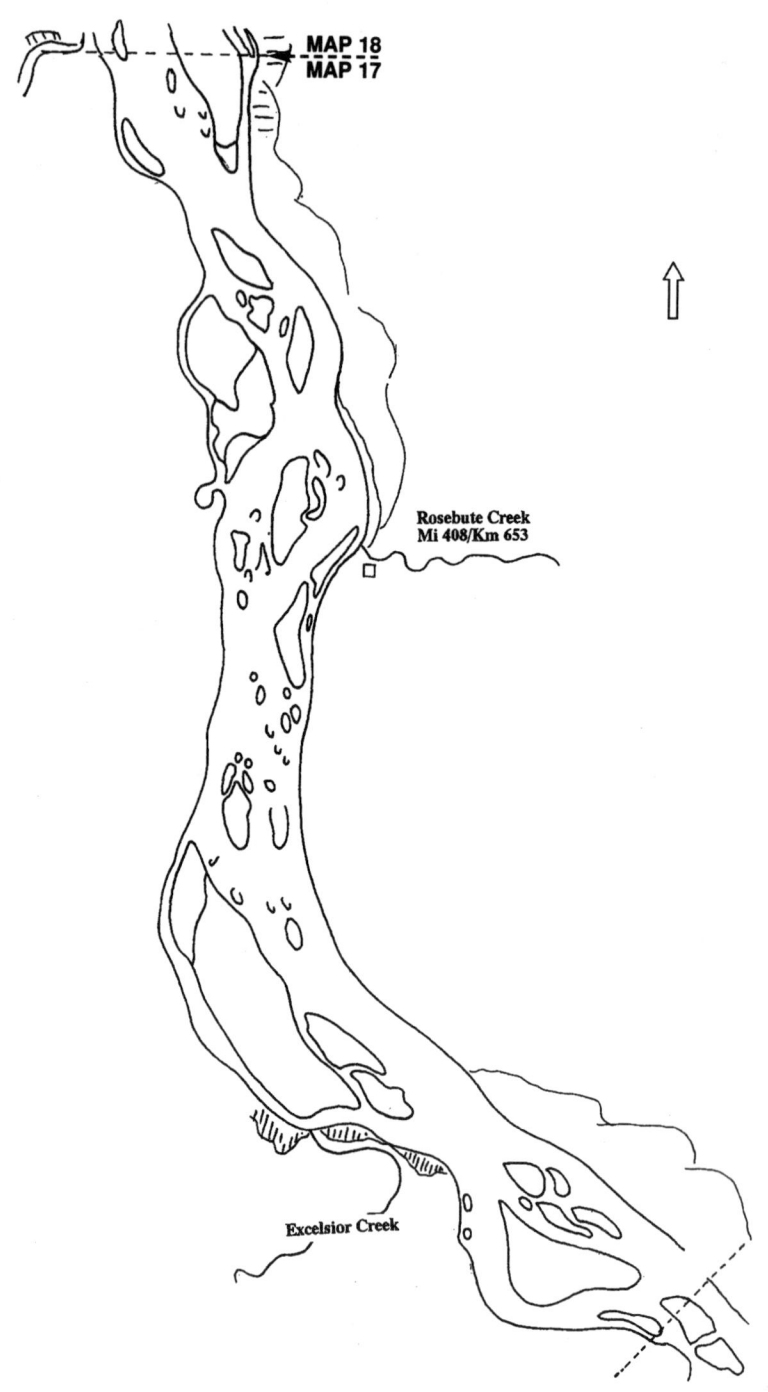

MAP 18
MAP 17

Rosebute Creek
Mi 408/Km 653

Excelsior Creek

MAP 17   **PAGE 44**

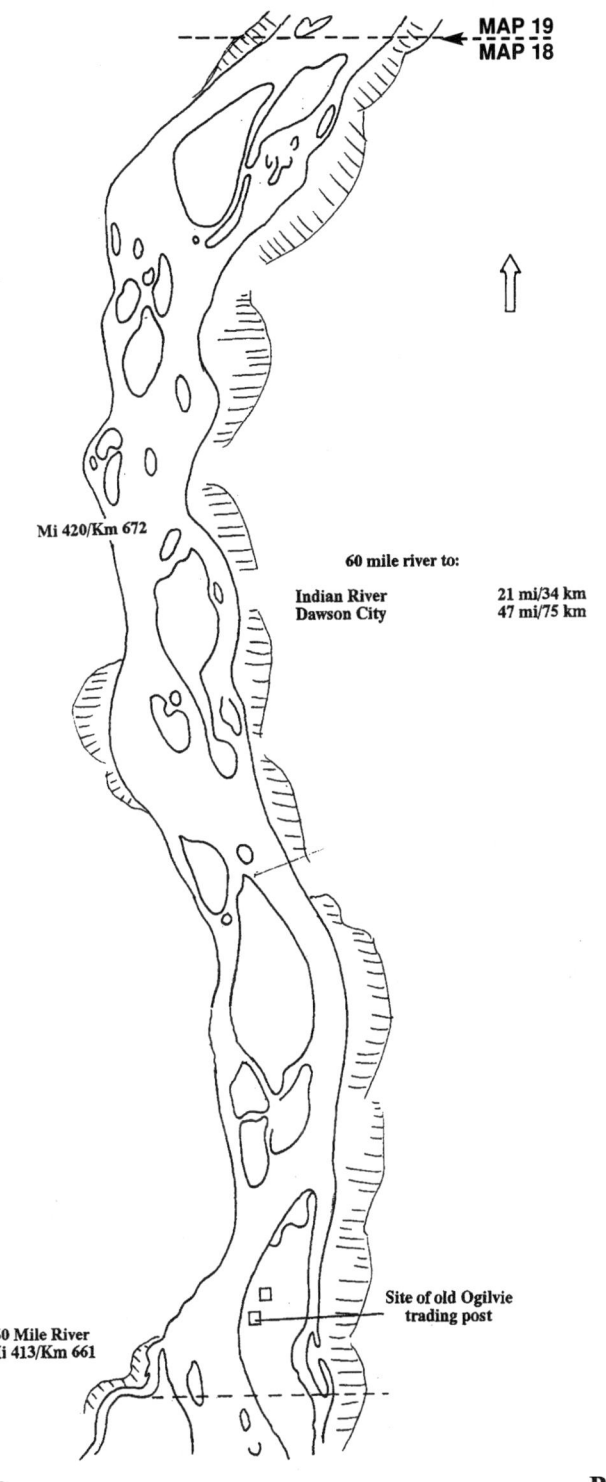

MAP 19
MAP 18

Mi 420/Km 672

60 mile river to:

Indian River   21 mi/34 km
Dawson City   47 mi/75 km

Site of old Ogilvie
trading post

43   60 Mile River
Mi 413/Km 661

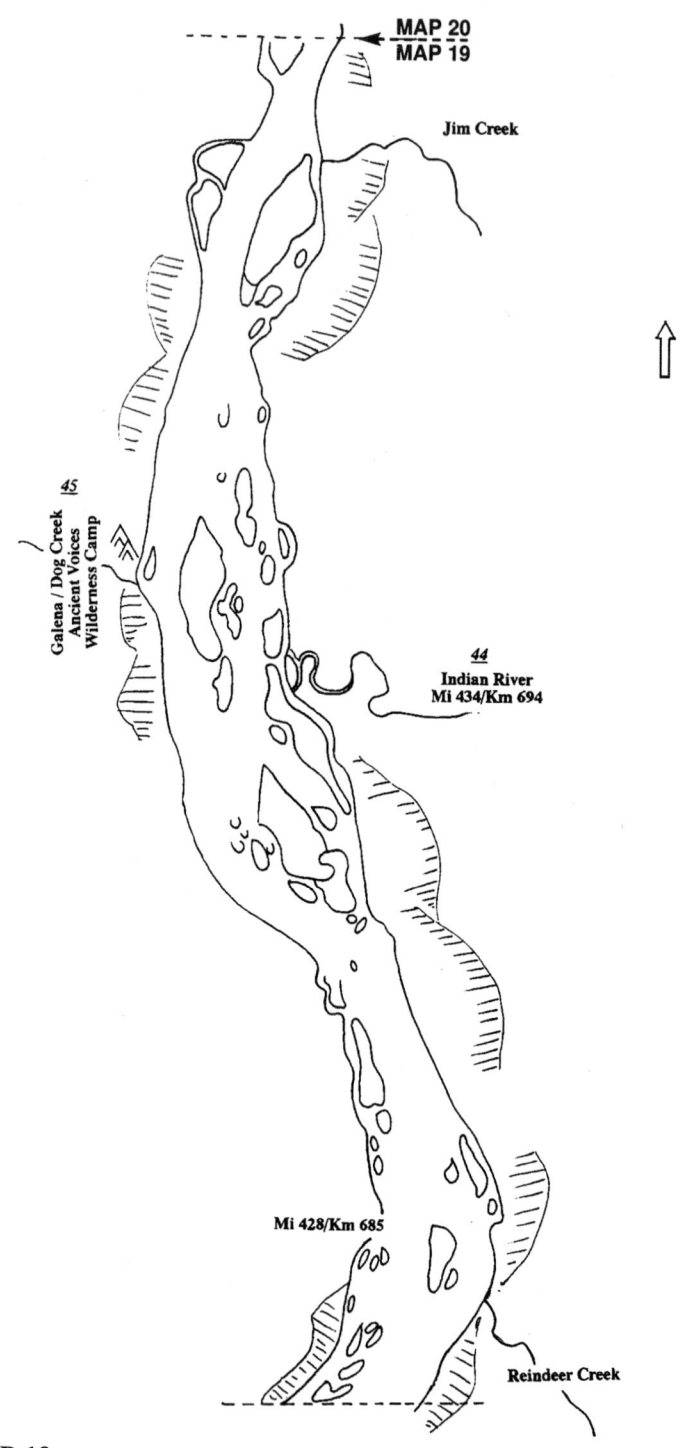

MAP 20
MAP 19

Jim Creek

_45_

Galena / Dog Creek
Ancient Voices
Wilderness Camp

_44_
Indian River
Mi 434/Km 694

Mi 428/Km 685

Reindeer Creek

MAP 19　　　　　　　　　　　　　　　　　**PAGE 46**

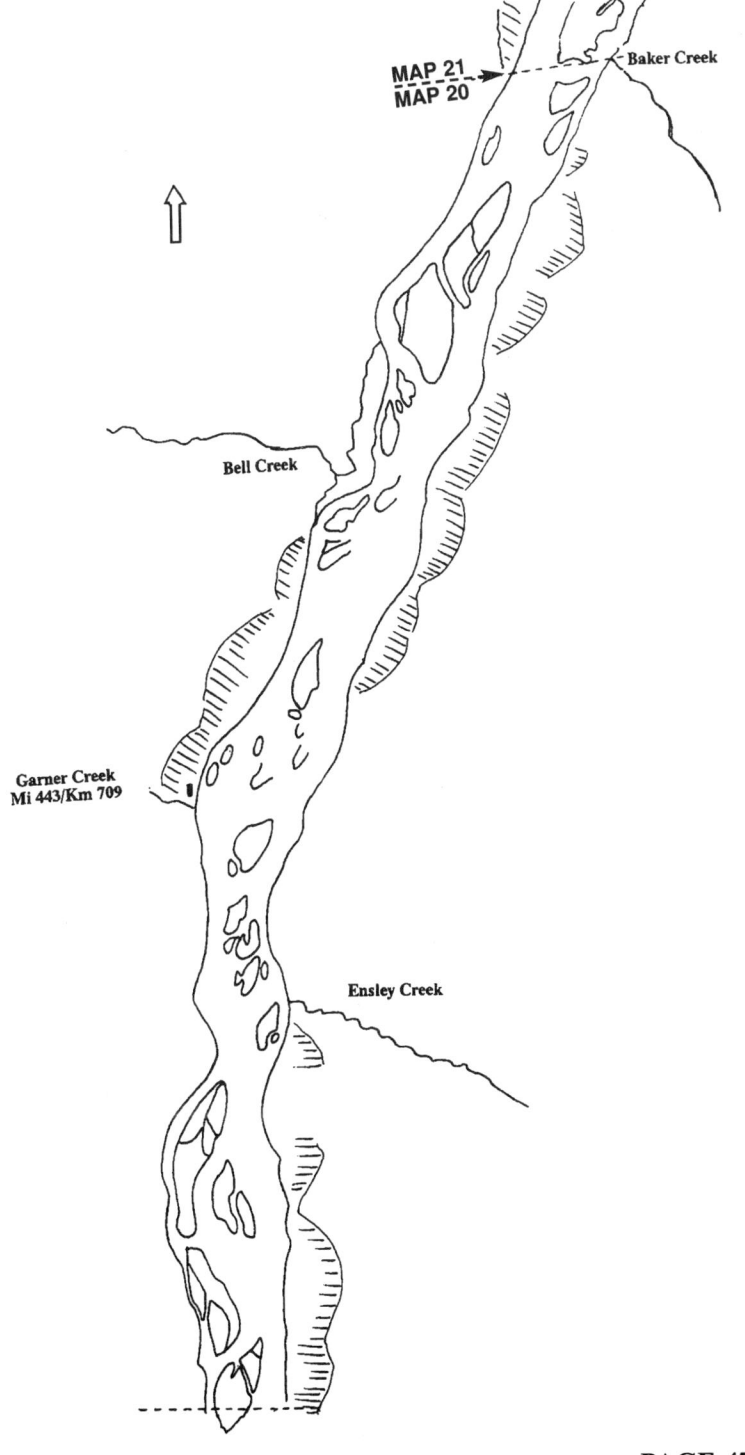

MAP 21
MAP 20

Baker Creek

Bell Creek

Garner Creek
Mi 443/Km 709

Ensley Creek

MAP 20

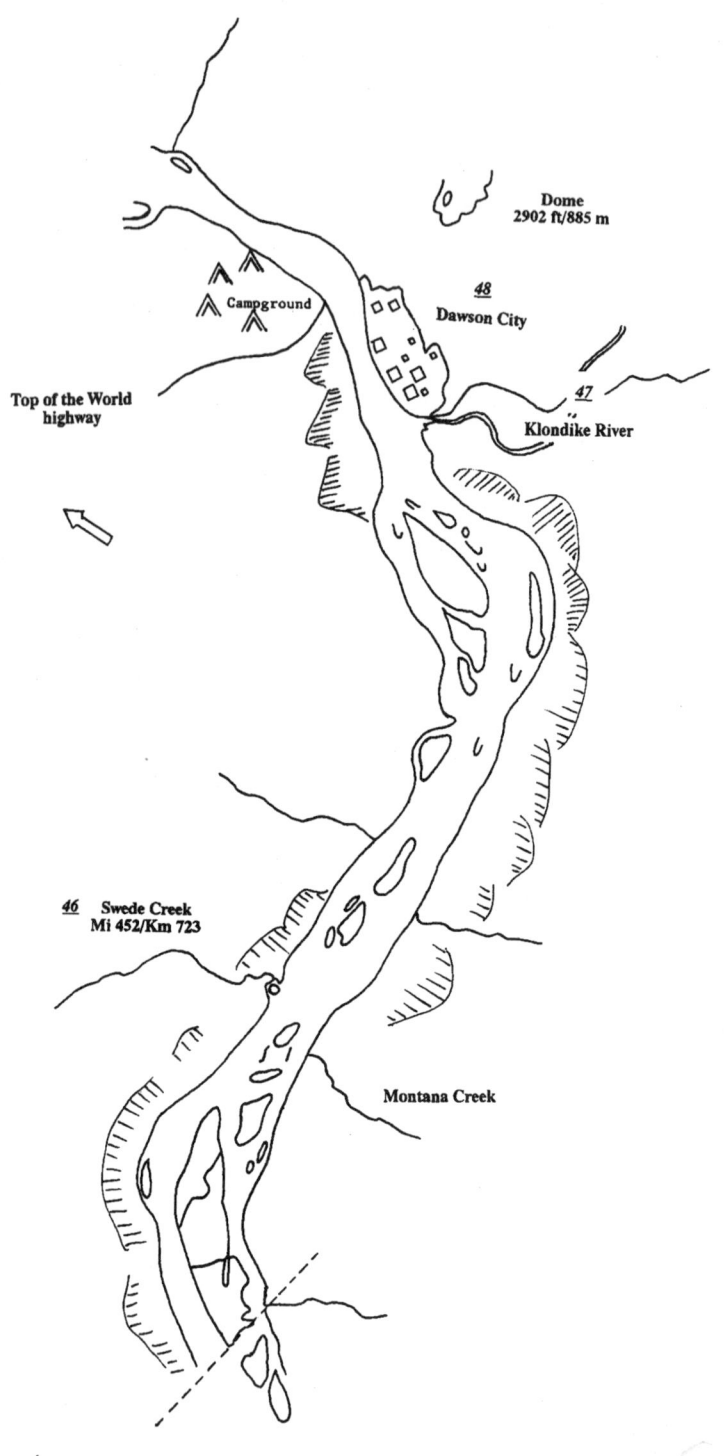

Dome
2902 ft/885 m

Campground

_48_
Dawson City

Top of the World
highway

_47_
Klondike River

_46_ Swede Creek
Mi 452/Km 723

Montana Creek

MAP 21

5

The Historical Log

1 The Village of Carmacks—Mile 200 (km 320)

MAP 1   **PAGE 28**

The Village of Carmacks is 110 road miles (176 km) north of Whitehorse and 200 miles (320 km) by river. It is 260 miles (416 km) by river to Dawson City. It is a frequent destination for canoe and boat traffic coming down the Yukon, Teslin and Big Salmon rivers. The Teslin River comes into the Yukon 110 miles (176 km) upstream at Hootalinqua and the Big Salmon River enters the Yukon 75 miles (120 km) upstream at the abandoned village of Big Salmon.

You can pitch your tent underneath the highway bridge spanning the Yukon River. The campground is used by both river and road traffic. Many canoeists try to arrive in Carmacks early in the day so they can continue on and camp further downstream away from civilization. Carmacks has most anything needed in terms of supplies and services.

The town is named for one of the codiscoverers of Klondike gold, George Washington Carmack. George was born to Hannah and Perry Carmack, September 24th, 1860, on a ranch in Contra Costa County, California. He first came north in 1882 as part of a U.S. Marine peacekeeping contingent stationed at Sitka, Alaska. In the fall of that year, the marines were relieved and George returned to the lower U.S.A. The north country, its people and the opportunities available, obviously made an impression on him. In the spring of 1885, he returned to Alaska, first entering the Yukon Territory via the Chilkoot Pass.

This was only five years after the first miners ever were allowed into the Upper Yukon River Valley. Prior to that, the Alaskan Chilkat Indians had effectively blocked access via the mountain passes for fear that the miners would interfere in their interior trade with the Yukon natives. While preparing for his trip into the interior, he was fortunate to have as mentors Hugh and Al Day. The brothers were French-Canadians who had struck it rich in the gold diggings in the Cassiar district of B.C. They were well known in the mining fraternity. More important, however, they had spent the previous year prospecting the Yukon River Valley as far north as present-day Dawson City. Their knowledge of the area was a great source of information and comfort to George and his men as they headed north into the little-known territory.

Between 1885 and 1896 George roamed the territory from the Chilkoot Pass to the village of Forty Mile on the Yukon River and as far north as Rampart House on the Porcupine River watershed, always on the lookout for that pay streak. Carmack and his Tagish Indian wife Kate spent some years in the early 1890s operating a trading post and working a coal seam that he had found at Tantalus. The mine workings are still visible on the hill overlooking the Yukon River and present-day Carmacks. In the summer of 1896, he and Kate abandoned both unprofitable ventures and headed downstream to the village of Forty Mile. Here gold had been discovered as early as 1886. That same year, George, with friends Skookum Jim Mason and Tagish Charlie discovered the Klondike gold.

The trading post and the Tantalus Coal Mine workings were the beginning of today's Village of Carmacks.

Just who was responsible for the first discovery of the Klondike gold has always been a matter of contention, particularly by one Robert Henderson, a prospector, who was in the area of the discovery at the time. Henderson was awarded a pension by the Canadian government in recognition of his efforts. At the same time, it is very apparent that gold was discovered in the vicinity several years before 1896, and the Forty Mile River just below present-day Dawson City had been yielding gold in paying quantities for almost a decade. It is also apparent that Carmack and his partners discovered the gold that initiated the Klondike gold rush.

The following item appears in a booklet written by George Carmack entitled *My Experiences in the Yukon*. It was printed by The Trade Printery of Seattle sometime in the 1920s. The booklet carries the following inscription:

Seattle December 29-1933
To Vancouver Public Library
From the Widow of the Discoverer of the Klondyke,
On Bonanza Aug 16th 1896.
Faithfully yours, Marguerite Carmack
- 1522 E. Jefferson St. Seattle, Wash.

The text of the booklet starts out with the following:

Resolution Offered by Seattle Lodge No. 2,
Yukon Order of Pioneers

Whereas, It has come to the knowledge of this Lodge that certain scurrilous articles, written by certain newspapers and magazines, have been circulated for the purpose of creating doubts as to Brother George W. Carmack's credibility.

Therefore, Be it Resolved, That we, the members of Seattle Lodge No. 2, Yukon Order of Pioneers, who knew him in those early and strenuous days of the Yukon, and who have associated with him more or less ever since that time, do hereby endorse him as a man of truth, honesty and integrity.

Whereas, The Grand Lodge of the Yukon Order of Pioneers at Dawson, Y.T. , has set aside the 17th day of August as "Discovery Day" in honor of Brother Carmack—a day that is celebrated with banquets and balls by every lodge of pioneers in the North and Northwest. We strive to perpetuate the memory of that day when the civilized world was electrified with the startling news that Geo. W. Carmack had found the combination to the lock on Canada's refrigerator, with the hope that that day (August 17th) may stimulate his brother pioneers in further efforts for the good of our wonderful Northland.

Geo. Carmack may never receive recognition from the Canadian Government for the days and years of nerve-racking hardships he endured in order that the trail might be made smooth and easy for civilization to follow. But Carmack is not a quitter. He is the type of pioneer who moves ever onward and allows no obstacle to bar his way. It is his nature to be a pathfinder. It is in his blood, inherited form his pioneer father and mother, with whom he crossed the plains to California in an ox team when he was a child. Plucky, self-reliant and physically fit, he would be an asset to any frontier country in which he elects to cast his lot.

Therefore, Be it Resolved, That we highly endorse Brother George Washington Carmack and recommend him as a loyal American citizen and very much of a man.

Dated May 2nd, 1922
F. E. KNOWLES
A. W. Arnaud, President
D. H. McDonald, Secretary

2 The Nordenskiold River—Mile 203 (km 325)

MAP 1   **PAGE 28**

The river was named by Frederick Schwatka in 1883 for Baron Nils Nordenskiöld, a Swedish Arctic explorer credited with being the first to follow the northeast polar route. In 1878–79 he sailed from the Atlantic Ocean eastward round the north coast of Siberia, through the Bering Sea and the Bering Strait into the Pacific.

The river is about seventy miles (112 km) long and its source is in the Miners Mountain Range to the south. An old trail leads from here, via Kusawa Lake, to the Alaska Panhandle. The Chilkat Indians used this route to reach the Yukon River Valley. "Thuch-En-Dituh," as the river was called by the Natives, literally translates to "meeting place." A large part of the old Indian trail became the Dalton Trail which used the Village of Carmacks as a terminus.

3 Shirtwaste Bend—Mile 206 (km 330)

MAP 1   **PAGE 28**

As mentioned earlier the curves that occur here are referred to as Shirtwaste Bend because of their resemblance to shirtails.

4 Murray Creek—Mile 212 (km 339)

MAP 1   **PAGE 28**

The creek was named for a Murray who located here at the turn of the century. It is also known locally as Myers Creek for Stephan and Marie Myer. They operated a number of wood camps along the river in the Carmacks area.

5 Mount Miller—2,850 Feet (867 m)

MAP 1   **PAGE 28**

This site was named for Captain Charles E. Miller. He arrived in the territory in 1897. He built the steamer *Reindeer* which operated on the Upper Yukon at the turn of the century. (Apparently the vessel had a large Reindeer Milk sign painted on the side which I believe was a common brand of condensed milk at the time.) Miller was an old mining hand from Pennsylvania and in the early 1900s his interests turned to coal mining. He developed the Tantalus Butte Coal Mine and operated this for a number of years. He sold the mine to the White Pass & Yukon Route company which operated a large number of river steamers.

6 Mount Monson—2,700 Feet (823 m)

MAP 1	PAGE 28

The mountain was named in 1911 for a constable of the Northwest Mounted Police. George Carmack in recounting the story of his famous gold find, mentions Monson as one of the three prospectors with Robert Henderson whom he met on the day of his find. The other two with Henderson being Swanson and Jack Dalton.

7 Tantalus Butte—2,568 Feet (783 m)

MAP 1	PAGE 28

The river winds so much here that it appears you are approaching this hill one moment and in the next view it appears behind you. The butte was named by Schwatka in 1883 after Tantalus, the legendary King of Lydia. He was condemned to stand up to his chin in water beneath fruit-laden boughs. Each time he tried to eat or drink, the water and the fruit would recede out of his reach.

George Carmack developed a coal mine here in 1893. Capt. Charles E. Miller of Mauchunk, Pennsylvania, took over in 1903 and over a period of two years mined some 40,000 tons of coal. The coal was burned in the boilers of the riverboats then operating on the Yukon River. More recently, the mine was reopened and for a time the Anvil Mine in Faro, Yukon, used the coal in their operation. It was again shut down in the late 1980s.

8 Five Fingers Coal Mine—Mile 218 (km 349)

MAP 1	PAGE 28

This mine was obviously named for the approaching rapids. It was also one of the properties George Carmack tried to develop prior to 1896. Charles Miller (see Tantalus Butte) also worked the mine. He apparently sank an initial slope (tunnel) of 350 feet. Miller sold the property to George Milton in 1903. Milton organized the U.S.-based Five Fingers Coal Company. The mine workings were abandoned by the company in 1908 apparently in favor of the Tantalus Butte mine.

Lignite coal seams were first reported here by George Mercer Dawson, a member of the Yukon expedition for the Geological Survey of Canada in 1887. Dawson reported finding coal seams from Five Finger Rapids (then referred to as Rink Rapids) to Tantalus Butte. In 1904, the Northwest Mounted Police annual report indicated that coal from this mine was tried as a fuel in their Dawson City buildings. The price and quality of the coal were not good and use of the coal as a stove fuel was abandoned, we assume in favor of wood.

9 Kellyville—Mile 220 (km 352)

MAP 2	PAGE 29

This old site is hard to find and is not to be confused with some new buildings in the area. Gerry and Rose Kelly lived here in the early 1900s. They built a large cabin at the site, ran a roadhouse and also managed to raise a large family. Rose was the granddaughter of Jack McQuesten, one of the original Yukon pioneers. Kelly Creek, a tributary of the Tatchun River, also appears to be named for the Kelly family. Prior to settling here, Gerry was the telegraph operator at Stewart River. This is presumably where he met Rose.

An old map, drawn in 1913, shows a sawmill at this location and labels the islands just below Kellyville the Reindeer Islands for the steamer that allegedly burned there in the winter of 1899–1900. The SS *Reindeer* was built by Capt. Charles E. Miller around 1897–98 when he was mining coal at Tantalus and Five Finger Rapids. The immediate vicinity around Kellyville was obviously a desirable place to build. Again referring to the turn-of-the-century river map, this location is marked as Wilsons Cabin which seems to indicate that perhaps he preceded the Rose and George as a resident here.

10 Five Finger Rapids—Mile 224 (km 358)

MAP 2	PAGE 29

This is probably the most photographed and best known landmark on the Upper Yukon River. Geologically, it is believed that a heavy bed of conglomerate rock originally blocked the river, creating a waterfall. In time the distinctive fingers were formed.

The rocks are very imposing, but historically vessels had no difficulty navigating the rapids. At high water some of the steamers had insufficient power to make the upstream passage. A cable house was built on the cliff to your right (east), just before you enter the chute of the rapids. Cables were strung out from here to the vessel coming upstream and a winch was used to supplement the steamers' own power.

It is believed that the rapids were named in 1862 by W. Moore of Tombstone, Arizona. The rock fingers are traditional nesting sites for a large number of gulls which have been calling the place home for some years. In 1883, Schwatka described their numbers as so many "that their taking to flight almost blocked out the sun." Schwatka also tried to rename these rapids "Rink Rapids." William Ogilvie of the Dominion Land Survey, disallowed this in favor of the local Five Fingers name in use by the miners. Ogilvie did apply the name Rink to the rapids five miles below.

In the early 1900s, the Department of Public Works and British Yukon Navigation Company, more commonly known as the White Pass, spent time and money widening the passage and blasting some of the larger rocks out of the channel.

3 accidents at Five Finger Rapids

2 canoeists missing in riv

By MASSEY PADGHAM
Staff Reporter

Carmacks RCMP are using a boat today in the search for two people presumed drowned as the result of separate accidents at Five Finger Ra Yukon River.

Three canoes have tipped over in the rapids, 30 kilometres north of C the last 10 days. RCMP say each canoe took the dangerous centre channe rapids and warned all canoists to use the safer right hand channel.

One person is missing following the overturning of a canoe yester Details of the accident are still sketchy, but the other person in the car safety, said RCMP Sgt. Ed Zawyrucha.

That canoe was one of two that overturned in the rapids yesterday. I cident, around noon, the single occupant of a canoe managed to make is believed to have lost his canoe.

In a similar accident June 16, a West German tourist in his 20s we partner hung onto the canoe and was rescued by RCMP using a helio Dennis Levy of the Whitehorse RCMP. Police had been tipped off by s ing the nearby Klondike Highway.

Police used a helicopter yesterday to search for the two m helicopter search was also made earlier for the victim of the June 16

RCMP say some of the people were likely not wearing life jackets added a reminder that canoeists should use them.

The Whitehorse Star, Thursday, June 25, 1981.

Bodies of drowned tourists recovered

The bodies of two men drowned in separate boating accidents have been recovered by the RCMP.

As a result of a boat search conducted by RCMP from the Teslin detachment on Saturday, the body of Jurjen Reiser was located at Jackson's Bay in the Brooks Brook area of Teslin Lake.

Reiser, a German tourist, went missing on the lake July 4 after a canoing accident.

Last week RCMP from the Dawson detachment found the body of an Oregon man drowned in the Five Finger rapids on the Yukon River. The body of Thomas Rob Farrell, missing since June 24, was found near Thistle Creek where the Stewart River enters the Yukon by another river traveller who reported the body to Dawson RCMP.

The channel you should use is the one on the extreme right! Canoeists have tried to use the center channels with disastrous results as is evidenced by the preceeding newspaper articles. At high water there are a number of standing waves in the passage, but these can be handled. Keep the canoe in the center of the channel and keep it moving. Several rocks are marked at the head of the second island below the rapids. Underwater obstructions do strange things to the water flow here and this area should be avoided by canoeists.

11 Tatchun Creek, Tatchun River —Mile 226 (km 362)

MAP 2	PAGE 29

George M. Dawson of the Geographical Survey retained this obviously Indian name during his survey in 1887. He did not obtain a translation. Topographical maps list this as the Tatchun River, locally it is known as Tatchun Creek. The creek has an annual run of salmon. In the spawning season, fishing is limited or restricted. Pay attention to the signs.

There is a public campground here with road access to the Klondike Highway.

12 Rink Rapids—Mile 230 (km 368)

MAP 2	PAGE 29

These rapids were named for Dr. Henry Rink of Christiana, Denmark, an authority on Greenland. Frederick Schwatka originally applied this name to Five Finger Rapids.

Stay to your right on the approach. As the rapids come into view, it will appear as if the entire width of the river is one white riffle. As you come closer, you will see a relatively calm channel open up along the right shore.

These rapids presented more of a problem to the steamers than the Five Fingers. The steamboat channel was through the middle left. Although deep in places, the rocks are numerous and several of the large steamers, including the SS *Dawson* and the SS *Casca*, came to their end in Rink Rapids.

13 Yukon Crossing—Mile 236 (km 378)

MAP 2   PAGE 29

This area is so named because the Whitehorse to Dawson City winter road crossed the river here. The remains of a large roadhouse and barn are still visible although the roadhouse has recently collapsed. The lodge, or roadhouse as they were more often called, had enough room to put up twenty-four people (two to a room). It also had a large kitchen/dining area and a small lobby/sitting room.

Summer traffic on the road was scarce but if you chose to use the road, you could cross the river here by a manually rowed ferry of sorts. Via the winter road, Yukon Crossing was 144 miles (230 km) from Whitehorse. If you were traveling the wagon road around breakup in the spring or freeze up in the fall, you could be stuck here for as much as ten to fifteen days waiting for the ice to be strong enough to cross on or for the river to clear of ice so that the "ferry" could be used.

Yukon Crossing is a pleasant place to camp. The highway is a little close and firewood can be a somewhat scarce, but there is room to move around.

On an old map drawn around the 1910s, the remarks "Old Slaughter Corral and Shack" are marked at Yukon Crossing. From this, I think it is reasonable to assume that cattle were slaughtered here and shipped downstream to Dawson City. It could also be that cattle were raised by the people at Yukon Crossing and sold as fresh beef to passing boats and camps in the area.

14 McGregor Creek—Mile 241 (km 386)

MAP 3   PAGE 30

We have no information on this spot. It has been suggested that it is named for a person that ran a trapline along the creek at the turn of the century. There are no signs of any permanent structures ever having existed at the creek.

15 Merrice Creek—Mile 241 (km 386)

MAP 3   PAGE 30

Merrice Creek is named for Homer Merrice. He lived here and also mined the creek in the late 1800s. During the topographical survey previous to the current one, the creek was mistakenly named Merritt Creek for Louis Merritt who owned the Williams Creek Copper Mine, located further downstream, from 1903–10.

The cabin remains at the site were built by prospector, trapper Afe Brown whose family still resides in the territory. Brown also operated a woodcamp on the large island just upstream. Brown, an American, went on a prospecting trip in the summer of 1944 and simply failed to return to his river home. The house remained empty until 1946. Walter Isreal, who had worked for Brown at the woodcamp upstream, moved into the house in the winter of 1946.

The Isreals at Carmacks, 1979.

Photo: Irene Pugh.

After Brown's disappearance, Walter Isreal worked at Brown's camp for a few years. He also had a number of other camps up and down the river until the late 1940s. Isreal and his wife lived at Merrice Creek until 1947. When river traffic came to a halt, the couple moved to Carmacks where they built and operated the Carmacks Hotel until his retirement.

More recently the cabins have been used by exploration crews working on nearby mineral deposits. The cabins offer little shelter and should not be counted upon for an overnight stay. The cleared area around the old homestead is a pleasant place to pitch your tent. The woodcamp cabin on the upstream island was dismantled and moved in 1936.

Boats had trouble pulling into Merrice Creek because of the shallow approach. Normally they would carry on to Williams Creek where the Isreals would have to call for their mail and supplies. This was probably why some people referred to Merrice as Williams Creek.

16 Williams Creek—Mile 243.5 (km 390)

MAP 3   **PAGE 30**

The activities in this stretch of river have always been centered around the Williams Creek Copper Mine. Williams was reputed to be the first on the creek and some of our research indicates that a Billy Williams was at the creek until the early 1900s looking for gold. This, however, has not been confirmed.

Leroy N. McQuesten (Jack), an early Yukon pioneer, writes that in 1885 on a river trip upstream to Fort Selkirk, he was accompanied by three companions. They were Thomas Williams, Mike Hess and Joe Ladue. McQuesten stopped at Fort Selkirk and the other three continued upstream to "check for minerals." This could also be the Williams who settled the creek for a few years, but again it is purely conjecture on my part.

From 1903 to 1910, Louis Merritt worked the copper mine with a crew of about eight to ten men. This activity was sufficient to warrant a small settlement at the mouth of the creek and also the building of a roadhouse some distance upstream. This, according to Walter Isreal, was "a single-story building with lots of rooms run by Mrs. Morrison."

Bornite is a copper ore which abounds in the vicinity. During the mining period in the early 1900s this spot also became known as Bornite City.

I have not been able to pinpoint any history for the period between 1910 and 1925 during which the mine was apparently taken over by three French-Canadians who remained there until the early 1930s.

Bornite City and Mrs. Morrison's roadhouse have disappeared into history and the only thing still visible are some old, rotten foundation logs and mining remains. In more recent years the mineral deposits have been explored by modern-day miners with modern equipment, and who knows, Bornite City may again come into being.

A substantial road runs through the bush behind Merrice and Williams creeks. Parts of this road can be seen further downstream. The road to the best of my knowledge goes from nowhere to nowhere, and has not been maintained during the last twenty years or more. Sections of the road have totally collapsed.

17 Hoo-Che-Koo Creek and Hoo-Che-Koo Bluff —Mile 247 (km 395)

MAP 3	PAGE 30

The name goes back to well before the gold rush. There are many interpretations of Hoo-Che-Koo but none are consistent. The name is obviously of Native origin, with the phonetic spelling perhaps slightly different between interpreters. Both the creek and the bluff were named prior to the turn of the century. Hoo-Che-Koo has been interpreted in association with "hootch," a vile home brew concocted by trappers, miners and other bush residents and used as a substitute for alcohol. It has also been associated with moose and with party gatherings.

Noted Yukon historian Allan Wright in his book *Prelude to Bonanza* notes as follows: "Reputedly, either an army deserter or an escaped convict from British Columbia - taught the Tlingits the technique of distilling liquor from molasses and sugar. None of the tribes adopted this art more eagerly than the Hootchinoo of Admiralty Island. Eventually the name 'hooch' applied to the product of all the stills, entered the language, immortalizing in a modest way the misdirected enthusiasm of these people."

George Dawson, when compiling his reports of his 1887 trip through the Yukon for the Geological Survey of Canada, shows Hoo-Che-Koo quite prominently. He brought back samples of the "copper-stained" portions of the rock, which when assayed contained minute traces of gold and silver. The silver content was assayed at .088 oz. per ton.

Hoo-Che-Koo may also be the white man's version of two Indian words. An early Indian dictionary compiled by Alexander Murray in 1847 lists Tsu-E and Tsu-Ko as an otter and a marten, both fur-bearing animals. Both animals may have inhabited the creek. The otter in particular was of special significance to the Indians. Native lore relates that the otter was considered a bad spirit. If caught in a trap, it should be left in the trap until the next day. A skinning knife used to skin the otter should never be used again but left driven into a special tree stump, never to be picked up again. The animal should be carried by one hind leg, a bit of prickly rose bush carried in the other hand would protect you from the bad spirits. Some believed that if you drowned you would turn into land-otter men. The otter persons were feared because they preyed on lost and drowning persons and would lure them to their village from which a person could never return and where he or she would eventually turn into an otter person.

It could well be that the creek has or had a resident family of otters and was so named so that it could be avoided. The bluff would be the landmark for the creek location.

The bluff was of special interest to early geologists who dated the rock into the Palaeozoic Age (500 million years old).

18 McCabe Creek—Mile 250 (km 400)

MAP 4	PAGE 31

The creek as named for old-time resident, trapper and trader, Thomas McCabe, who operated a post here in the early 1900s. McCabe Creek post became the area post office after the closing of Fort Selkirk Post Office in 1952–3. I have marked the old Whitehorse–Dawson road on the map. It intersected the creek at this point.

In time, as the Klondike Highway was built, McCabe's post became known as Midway Lodge. It was roughly halfway between Whitehorse and Dawson City. The lodge burned in 1995.

McCabe was apparently quite a gardener and was renowned for his annual potato crop. The gardens are still used by present-day owners of the property. A house and fuel storage facilities have recently been built on the river bank at McCabe Creek. Several shallow draft river-boats are being used to transport fuel from here to Ballarat Creek at mile 344 (km 550) downstream.

The prominent rock on the left bank, just past McCabe Creek for some reason appears on the old maps as Rhinoceros Rock. Perhaps someone with a very active imagination thought it looked like one.

19 Minto—Mile 258 (km 413)

MAP 4	PAGE 31

This is the last place for highway access until you reach Dawson City. Minto was originally an Indian settlement named Kitl-Ah-Gon. Translated this meant "the place between the high hills." It was renamed Minto in 1900 in honor of a visit by Lord Minto the governor general of Canada. Some of the older river maps show an "Old Minto" slightly upstream from the present-day sight. Recently there has been some clearing done at this old site and a large river barge ties up there.

Minto was a stage stop on the old Whitehorse to Dawson winter road and quite a large village in its heyday. It was a major staging area for the Klondike Highway construction crews and for several years during the road to river transition, freight was landed here for the towns of Elsa and Mayo on the Stewart River and of course for Dawson City. For this reason in some history books it is also referred to as Minto Landing.

There is little of the original village left. It has been turned into a public campsite which is in constant use by road travelers. Unless you feel like socializing, it is not a good place for an overnight stay.

In recent years, on a spasmodic basis, several barges have been operating out of here for destinations downstream. If you happen to meet a barge on the river, keep in mind that the pilot must stay in a specific channel and you should give way. The wake of the barge is quite substantial. If meeting the vessel take capsizing precautions.

Garbage can be disposed of in the Minto campground containers.

20 Von Wilczek Creek—Mile 260 (km 416)

MAP 5	PAGE 32

This was named by Schwatka in 1883 for Graf Von Wilczek of Vienna, Austria. He originally designated "river" status to the stream. Geographical survey parties later reclassified it to a creek.

21 Big Creek—Mile 262 (km 419)

MAP 5	PAGE 32

The creek was named by an ex-Yukon Field Force soldier John

McMartin who settled here after taking his discharge from the army at Fort Selkirk in 1899.

A popular fable involving this creek suggests that a large amount of gold stolen from Dawson City is buried here. The thief was apprehended at Big Creek but the gold was never recovered. The source of this story has never been identified nor a parallel occurrence in Dawson City substantiated.

McMartin, like almost everyone living along the river, supplemented his income by operating a wood yard. Old river charts indicate that he had a wood pile just down from Big Creek. On some of the turn-of-the-century maps, this creek is also marked as Black Creek.

Miners are currently operating in the area and there is no opportunity for a camp.

22 Devil's Crossing/Hell's Gate
—Mile 267 to 273 (km 427–437)

MAP 5   **PAGE 32**

This was, and still is, one of the worst areas for powerboats to navigate. Devil's Crossing is where the river steamers crossed over to the east side of the river (going downstream). Hell's Gate is always shallow, rocky, narrow and fast. During the steamboat era it was not unusual to have several of the large steamers stuck in this stretch of river at the same time, particularly if a "meet" was necessary. This was a sitiuation where an upstream boat met a vessel coming down. To alleviate some of the problems, a wing dam was built against the cliffs to divert some of the fast water into the channel and to wash some of the gravel bars out of the "gate." Iron rings were driven into the rock bluff of the gate so that steamers coming upstream could use them for an anchor and winch themselves through by using their own foredeck winch machinery. Some of these rings are still there. Watch for them as you are about halfway through the gate.

Since the discontinuance of paddlewheel steamers in the 1950s, this stretch of river has no longer been maintained, and for motorized river traffic it is again becoming a real problem to get through. The pilots of the barges running out of Minto will sometimes mark a channel in the gate. Markers consist of a brightly colored plastic bottle attached to a weight. Please do not disturb these markers.

The hillsides around Hell's Gate have been burned badly during the last few years. One bonus is the abundant fireweed that covers the hills as a result. If you go through on a clear, sunny day, have your camera ready.

23 Ingersoll Islands—Mile 269 to 277 (km 430–443)

MAP 6	PAGE 33

These were named by Schwatka in 1883 for Col. Ingersoll, U.S. Army, Washington, D.C.

24 Wolverine Creek—Mile 274.5 (km 439)

MAP 6	PAGE 33

Wolverine Creek is easily recognized. A distinctive rock bluff borders the creek mouth. In recent years the mouth of the creek has become very shallow, and any approach by powerboat should be made cautiously.

The creek is obviously named for the animal which probably inhabited the creek area in years past. The wolverine, Lekh-Ethu-E in the Kutchin Indian language, has been credited with almost superhuman qualities. The Natives considered it a destructive and devil-like superhuman spirit. Out of pure devilment, it will systematically destroy all the animals caught in a trap before the trapper can get to them. It will continue to do so until it is killed by the trapper or the trapper moves. Native lore also recognizes the animal's great strength. To give a dog stamina and endurance it is pulled through an inside-out wolverine skin as a pup. For the same reason, babies were sometimes accorded like treatment. Wolverine fur does not frost up easily in cold weather and therefore it is favored for parka trim.

It is generally believed that the early Russian Traders did not travel any further upstream on the Yukon River than Fort Yukon at the confluence of the Porcupine River. There have been signs found at Wolverine Creek which seem to indicate that an early Russian exploration party visited here long before Robert Campbell's discovery of the Yukon River. However, these are unconfirmed reports.

64

Wolverine Creek. *Photo: Gus Karpes.*

25 Grenier's Coal Mine—Mile 275.5 (km 441)

MAP 6   **PAGE 33**

A small deposit of coal was discovered here in 1910. George H. Grenier was the original operator of these diggings. He hand-mined the property in summer and hauled the coal into Fort Selkirk by wagon.

26 Slaughterhouse Slough—Mile 280 (km 448)

MAP 6   **PAGE 33**

Back in the days of unrefrigerated transportation, one enterprising individual drove a herd of beef cattle to this point, presumably on the old wagon road. From here he felt that the herd could be shipped as fresh beef the rest of the distance Dawson City without spoiling. The animals were butchered and shipped by barge. This slough has forevermore been known as Slaughterhouse Slough.

27 Pelly River—Mile 281 (km 450)

MAP 6   **PAGE 33**

The Pelly River was named by Robert Campbell in 1840 for Sir John Henry Pelly 1777–1852. Pelly was the governor of the Hudson's Bay Company. Campbell came in to the Pelly River about 335 miles (539 km) upstream where he had built a trading post which he called Pelly Banks. In 1843, Campbell descended the Pelly River into the Yukon River Valley. Upon seeing the Yukon River, he felt that the Pelly River was in fact the larger of the two, quite understandable when you see the Yukon and Pelly side by side at Fort Selkirk. He named the Yukon River that you have traveled so far, the Lewes River and continued the name Pelly for that part of the Yukon River below the confluence.

In 1848, Campbell built the original Fort Selkirk at the mouth of the Pelly River on the upstream, north bank of the Yukon. He moved the fort across the river to its present location in 1852. Nothing remains at the original site.

The old Whitehorse to Dawson City wagon road has also been marked on the map. The roadhouse/stage stop was part of a turn-of-the-century farm. This was a busy place in winter until the road was rerouted to a location near present-day Pelly Crossing on the Klondike Highway.

Wagons and sleds were used to cross the frozen river. In late spring when the ice became rotten and dangerous, the wagons and teams were positioned to operate between river crossings. To get the mail and other priority cargo across, it was loaded into a canoe which in turn was loaded onto a sled. The sled was then pulled across the ice by a crew who reasoned, that if the ice broke, the canoe would drift free and they could save themselves by hanging onto it or clambering aboard. This was a little hazardous, but there is no record or memory of any fatal incidents involving this practice.

28 Fort Selkirk—Mile 282 (km 451)

MAP 6	PAGE 33

The site of Fort Selkirk is located on a high bank just below the confluence of two major rivers, the Pelly and the Yukon. This strategic location made it an ideal meeting place for people throughout the centuries and it is generally believed that the location was the scene of many meetings and trading sessions between the Indian people of Selkirk, the surrounding area and the Chilkat Indians from the coast of Alaska.

This was also the site of the first non-Native settlement on the Upper Yukon River. Fort Selkirk was built at this location by Robert Campbell of the Hudson's Bay Company in 1852, the same year that it was sacked by the Chilkat Indians.

Campbell had been here twice before. In 1843, with a crew of six he reached the confluence on an expedition out of Pelly Banks, a

Hudson's Bay Post about 335 miles (539 km) upstream on the Pelly River. On the second trip, in 1848, he constructed the original Fort Selkirk across the river at the confluence of the two rivers. During his time with the Hudson's Bay Company and while taking these trips, Campbell kept meticulous diaries. Unfortunately these were destroyed by a fire in 1882. We do have the benefit of his memoirs which were written around 1892. The following are excerpts from the memoirs.

On the 6th day (believed to be June 16th, 1843) we reached the junction of a large river flowing from the S.W. which I named the Lewes after J. Lee Lewes. There we camped for the night. Early next day a short distance below the forks, we came upon a large band of 'Wood' Indians, whom we took completely by surprise, which almost amounted to awe, as they had never seen white men before.

Campbell wished to continue his exploration further downstream but turned around and went back up the Pelly River. His memoirs provide the reason for this decision.

Two of their leading chiefs, father & son, named Thlin-ikik-thling and Hanan, were tall, stalwart, good-looking men, clad from head to foot in dressed deer skins, ornamented with beads & porcupine quills of all colors. They said that inhabiting the lower river were many tribes of bad Indians, who would not only kill us but eat us. We would never return, & our friends coming after we would unjustly blame them. All this frightened my men so much that I had reluctantly to consent to turn back, which perhaps under the circumstances was the best thing we could do, as we were not equipped for a longer trip; and I learned afterwards that it would have been madness to proceed, unprepared as we were for such an enterprise.

On June 1, 1848, Campbell and crew returned and started building the original Fort Selkirk on the opposite, upstream shore to the present-day site.

It was in early August that year that Campbell first became acquainted with the Chilkat Indians who where four years later responsible for the sacking of Fort Selkirk across the way. Again, I quote from his memoirs.

One evening in early August, when a good many of these local Indians were about us, we heard a noise and singing & shouting up the Lewes. The Indians explained to us as best as they could who the strangers were (they were Chilcats) & advised us to hide our working tools & everything movable unless we wished to have them stolen by the strangers who where adept at pilfering.

Campbell describes the visitors.

The Chilcats belong to the Coast Indians along Lynn Canal, who had long carried on a bartering trade with the Indians in that quarter; & the only articles from the outside world—indifferent though they were in quality—these poor Indians had obtained in trade [came] from the Chilcats. I may add that such a thing as fair dealing was unknown among the Chilcats, whose motto was 'might is right,' & who were civil only when they were the weaker party.

From 1848 to 1952 the Hudson's Bay Company crew occupied the site at the mouth of the Pelly River. Campbell was not always present and left James Stewart, his assistant, in charge of the fort. In 1851, Campbell drifted the river from Fort Selkirk to Fort Yukon or Youcon House as it was then called. The Hudson's Bay Company fort at the confluence of the Porcupine and Yukon rivers was founded by John Bell in 1843. During the winter of 1851 and spring of 1852, Campbell started to move the fort to its present location. In a letter to James Anderson, the Hudson's Bay official overseeing the district, he explained the move. He said that the new site had actually been chosen on their arrival at the Forks in 1848. He explains that at that time being doubtful of his reception and the disposition of the Indians, who were numerous, he instead built the fort at the confluence of the two rivers.

In the summer of 1852, the first real "outfit" arrived via Lapierre House and Fort Yukon, and Campbell was looking forward to a busy and profitable winter. This was not to be.

The pillaging of Fort Selkirk was spectacular only in the sense that it was the only such incident in the recorded history of the Yukon Territory. It began on Friday, August 20, 1852. Campbell and three helpers were at the site of the old fort cutting hay when a large party of Chilkat Indians came drifting down the Lewes River. He and his crew quickly dropped everything and headed for the other side of the river but some of the Chilkat Indians beat them to the fort.

For three days the Chilkats terrorized the occupants of the fort. They systematically destroyed everything that they could not take with them. Campbell was left with insufficient trade goods and supplies to see him through the winter and he had no choice but to drift downstream to the post at Fort Yukon. He was never to return.

It is a wonder that there no one was killed in the incident. In the letter to Anderson on November 4th, 1852, Campbell explains the episode further.

An instantaneous rush was made upon me, with their guns and knives. Others seized me by the arms. Two of the guns snapped. One Indian as he sprung at me with a knife, ripped up the side of a dog that came across him, and the blood off the blade crimsoned my arm as I evaded the blow. In one of the guns aimed at me (a brass blunderbuss) I saw four bullets put a little before the fray began. My pistols, which were concealed in my belt, were wrenched from me before I could fire; in fact an attempt to do so would have been in vain, and could have ended only in the indiscriminate murder of all.

He continues describing the assault and his escape downstream by boat. He returned to the fort on August 23rd accompanied by a number of friendly Indians he had encountered downstream but found the Chilkats had left. The letter continues.

"I regret to say that except ammunition and tobacco, but little else of the entire outfit has been traded. Not a grain of powder or rag of clothing was left. Cassettes, dressing cases, writing desks, kegs and musical instruments were smashed into a thousand atoms and the house and store strewed with the wreck, a sight to madden a saint."

The forceful eviction of Campbell and his crew had the desired result as far as the Indians were concerned as it was not until many years later that the Hudson's Bay Company again entered the Yukon trade.

In 1889, Arthur Harper, an early Yukon pioneer, started a trading post at "Campbell's Fort" as the place was then known. Harper operated the post here until the Klondike gold rush started.

St. Andrew's Anglican Mission was established at Fort Selkirk in 1892. The Catholic St. Francis Xavier Church appeared later in 1898. The present St. Andrew's Church was built in 1929. Both church buildings stand today and are worth a visit.

The Dalton Trail was extended to Fort Selkirk in 1897. People, cattle and goods came in on the trail and traveled downstream to Dawson City from here. For a time, Fort Selkirk was considered as the capital of the Yukon Territory which honor was bestowed on the new City of Dawson in 1897. That year, Dawson City already boasted a population of 5,000 which was to swell to 30,000 and as much as 40,000 people by the summer of 1898.

In 1898, the townsite was surveyed and organized into lots and blocks. The Northwest Mounted Police built a post and storehouse and the Yukon Field Force arrived. This was a force of some 200 officers and men of the Canadian Army, who were brought into the country.

The Yukon Field Force, a unit of the Canadian militia, existed for only two years. The force was created and posted to this remote part of Canada to, first of all, present a symbol of Canadian sovereignty and, secondly, assist and aid any civil power as might be required in the turbulent atmosphere of a gold rush.

In 1899, with the completion of the Whitehorse to Dawson City telegraph line, a telegraph station was built and a permanent communication link with the outside was established. In these early days, to provide entertainment during the long winter hours, frequent dances were held alternatively at the roadhouse across the river or at the post itself. Merry makers would travel to and fro across the frozen river by dogsled. Music for the festivities was provided by a gramophone belonging to the trading post and the dancing would go on until the small hours of the morning. One of the more popular dances was a dance called the Lancer. This was a regimental version of today's square dance. White women were scarce and the local Native girls were often called upon to increase the number of dancing partners.

Things started to slow down in 1900. A number of Yukon Field Force members had already been transferred to Dawson City and in 1900 the force was disbanded. Some of its members remained in the area as civilians but most left the north for good.

The town had grown so much that in the year 1900, accommodation could be obtained at the Selkirk Hotel, the Savoy Hotel, the Seattle Hotel and the Hotel Francais, but unfortunately the population was already on the decline and commerce in Fort Selkirk was also on the wane. The Northwest Mounted Police detachment closed down in 1911.

In the 1920s and 1930s, the population steadied at about twenty-five whites and 200 Natives. St. Andrew's Church was built in 1929 and the RCMP detachment was reestablished in 1932. In 1936, the Hudson's Bay Company returned to town and this time managed to stay for sixteen years until June, 1951, when the store and the manager's residence were moved to Fort Nelson, B. C.

The completion of the Klondike Highway in the early 1950s spelled the death of Fort Selkirk. The Selkirk Indians moved almost en masse to the highway community of Pelly Crossing, thirty-six miles (58 km) upstream on the Pelly River. The RCMP detachment and post office were closed and for many years the only winter residents were Selkirk natives Danny Roberts and his wife Abigail.

During the 1980s some selective refurbishing was done in an effort to preserve some of the old buildings. More recently the Selkirk First Nation people, descendants of Campbell's first trading partners have taken over in the rebuilding of the site.

Fort Selkirk.   *Photos: Gus Karpes.*   School house at Fort Selkirk.

Camping at Fort Selkirk

This is one of the few places along the river where you are required to camp at a specific site. The campsite is not hard to find and is well marked. Aim for the flag pole. The camping restriction may come as a bit of a shock after the freedom of the past week. You do have the option of timing the visit to Fort Selkirk during the day and continuing on. A stop at this historic settlement is a must and is one of the highlights of the trip. You should plan on spending the better part of a day there.

29 Old Wood Camp—Mile 292.5 (km 468)
Marshall's Ranch—Mile 304 (km 486)

MAP 7, 8   PAGE 34, 35

Both these camps supplied great quantities of cordwood for Fort Selkirk and for the steamers that traveled the river in summer. For the steamers, a desirable wood camp to stop at for refueling was a camp that offered "seasoned" wood. This meant that the wood was cut at least one full year before it was used in the boilers. Horses, wagons and sleighs were used to bring the cut wood to the river bank were it was stored for easy access by river traffic. Many a camp operator lost

71

his entire winter inventory by picking the wrong location for the pile. At breakup, an ice jam could flood the area and wash the entire supply of firewood downriver, scattering it over miles of country.

One of the more tragic stories along the river takes place here in 1941. One camp was owned by a fellow named Menzies who also had his family living and working with him. The other was owned and operated by an elderly fellow named Marshall. Both had been in the wood business for some years and were always competing with one another. No socializing went on between the two camps and for some unknown reason, neither was welcome at the other's camp.

Laurent Cyr (see Britannia Island) hiked from his camp to Fort Selkirk in April, 1941. He stopped at Marshall's camp and found him totally incapacitated by gout. Both his feet were a frightful mess, swollen to twice their normal size and, of course, very painful. When Laurent arrived, Marshall was totally out of drinking water. He had crawled to the door and window through which openings he had been scooping snow. He used a long-handled spoon and of course he had scooped up dirt and everything else with it to where it was probably doing him more harm than good.

Laurent knew that he could do nothing for the man's illness. He filled up Marshall's water buckets, left them to hand and promised that he would stop at Menzies' camp to report his illness and also report to the RCMP at Fort Selkirk. Marshall, even in his shape, still protested and refused to consider help from the Menzies camp downstream.

It was spring and travel along the river was extremely hazardous. Laurent had to contend with open water, mountains of jammed ice and open creeks. Nevertheless, he got to Fort Selkirk in record time and reported Marshall's condition to the RCMP constable before stepping onto the aircraft for his trip back to Whitehorse.

Boating the river was out of the question until break up. Weather and ice conditions were dangerous and no one could go to Marshall's assistance until the river ice broke. The first boat to stop at his place found Marshall's body. He had been dead for some time. The sad part of the story of course is the cantankerous attitude of the two camp operators. Their personal animosities interfering to the point of caus-ing the unnecessary death of Marshall.

I have also marked several locations with the name Beaton. Finlay Beaton was a busy wood camp operator that lived in the area in the early 1900s, possibly being the builder of the Marshall Ranch and wood yard.

30 Selwyn River—Mile 317 (km 507)

MAP 9   PAGE 36

This river was named by Frederick Schwatka in 1883 for Doctor Alfred Richard Cecil Selwyn of England. Dr. Selwyn was born in 1842 and during his lifetime worked for the Geological Survey of Britain, Australia and Canada. He retired as the Director of the Geological Survey of Canada in 1895.

The Selwyn is short and has its headwaters in the Dawson Range of mountains about twenty-four miles (38 km) from here. The river was originally prospected by a man named Duval, an ex-jailer from Tacoma, Washington. In August, 1898, he staked a discovery claim which he shortly thereafter sold. Little gold was found and it was afterwards asserted that he had salted his claim (from *Yukon Places and Names* by R. Coutts).

Hayes Creek is a tributary of the Selwyn River. Klines Gulch is a tributary of Hayes Creek. R. Coutts, who supplied the above information on Duval also tells the tale of Kline, after whom the gulch was named. In 1904, he allegedly found the only gold ever located in the Selwyn watershed. Kline later committed suicide in a most accommodating manner. He set two charges of dynamite. The first to blow himself up, the second to collapse the walls and sod roof of his cabin on top of himself thereby effectively burying himself under three feet of soil and debris.

The Northwest Mounted Police had a post here in 1898 and one year later a telegraph station was opened. Although the police post closed in 1905, Selwyn became a regular fueling stop for the river steamers.

Selwyn has never been one of my favorite stops. The undergrowth is thick, the mosquito population always seems to be extremely hungry and the surrounding bush seems to emit a moldy, almost tropical odor. This seems heightened by the inherent lack of sunlight at the old site. The cabin remains are overgrown and difficult to find. They are surrounded by a good supply of berries. These are probably hybrids that originated with the domestic berries imported by the past residents. Several times while there, I distinctly felt I was robbing some bear's larder. The smell of a summer bear is quite pungent and distinct. This smell and the animal droppings around the berry patch convinced me to depart and I have been giving it a pass ever since.

Not a good place to stay.

31 Isaac Creek—Mile 328 (km 525)

MAP 10   PAGE 37

There are two Isaac Creeks in the territory, the other flowing into Sekulman Lake west of here. Isaac was the name of a well-known Indian family of the Aishihik band (pronounced aziak), who roamed the country with Jack Dalton in the 1890s. It is assumed that this creek was named for a member of that family.

Idaho Creek flows into Isaac Creek about five miles upstream. Both creeks were prospected and worked extensively at the turn of the century. The cabin remains at Isaac Creek are of a cabin originally built in the Minto area. The woodcamp operator that worked here in the 1930s moved in from Minto and brought his living quarters with him.

Isaac Creek is also the destination of equipment, fuel and supplies that the barge out of Minto drops there. This supplies Casino Camp— an exploration and mining camp some distance inland.

Camping at Isaac Creek is not the greatest for the same reason as Selwyn. We would recommend that you carry on and look for a better site.

32 Britannia Creek—Mile 335 (km 536)

MAP 11   PAGE 38

The "discovery" claims on this creek were staked in 1911 by E. L. de la Pole and C. M. Printz who named the creek. It has seen mining activity on and off since then. Second world war vintage equipment lies desolate on the riverbank as a reminder of past ambitions. This is not a bad place to camp and there is a fairly large cleared area close to the river.

Britannia and Canadian Creek, a tributary about six miles away, are currently being mined. The placer operations are some distance away, but on a quiet evening you can probably hear some heavy equipment being used in the mining work.

33 Britannia Island—Mile 336 (km 538)
(A woodcutter's lament)

MAP 11   PAGE 38

This is the local name given to the first large island below Britannia

Creek. It is one of the many unmarked locations where a semipermanent woodcamp was operated during the reign of the Yukon paddlewheelers.

It was common for the summer boat crews to occupy their winters cutting wood. Many had no other winter occupation to which to turn. The normal procedure was to contact vessel owners in advance which meant making a deal with the British Yukon Navigation Company, more commonly known as White Pass & Yukon Route, as they operated most if not all of the large vessels on the river. They would agree on a likely location for a woodpile and fueling stop, a price per cord of wood and the amount of timber required. The wood contractor would then go into the bush to do the cutting himself, or perhaps hire others to go in for the winter on a contract per cord basis. A cord of wood is a cubic measurement of four feet by four feet by eight feet (128 cubic feet).

On October 7th, 1940, the steamer *Casca* dropped her gangplank on the island to deliver two energetic Whitehorse individuals, Laurent Cyr and Murray Stephens. They carried limited supplies, a little bacon, beans and macaroni and prepared to spend the winter here cutting wood. They had been told that the country had an abundant supply of wild game and that it would be easy to supplement their meager rations with a .30-30 rifle which was included in their kit. By prior agreement they would live in separate cabins and would each cut their own pile. They utilized an existing cabin at the head of the island and built Murray a second cabin about one mile downstream.

The contractor was Brod Cyr, Laurent's brother. The boys were to receive $3.75 per cord for their efforts. It was to be cut and stacked on the riverbank where the boats could pick it up easily during the summer months. Trees were cut, limbed and left in the bush by the boys. They then hired Walter Isreal, another rivercamp operator to haul the logs onto the beach for them at $1.00 per cord. Isreal apparently was the only one in the area with horses.

During the winter months, Cyr and Stephens were pretty much left to their own devices and Cyr could not recall getting any visitors. He snowshoed into Coffee Creek several times to pick up their mail. The lack of a social life did not really bother the boys as much as the total absence of any game. This meant a steady, boring diet of macaroni and beans. Not even so much as a squirrel was spotted and nary a shot was fired all winter.

As spring came it was obvious that the remaining victuals were inadequate for the two of them and in mid-April, Laurent chose to be the one to leave and started his snowshoe trek to Fort Selkirk. From here, he intended to catch the airplane home to Whitehorse. Laurent felt that he had about 250 cords of wood cut.

When he returned to Whitehorse, Laurent and his brother where advised that they would not be paid for the wood until it was picked up by the boats and burned. As the wood was green, the boats would not pick it up until the summer of 1942. Although about thirty cords were picked up during the summer of 1941, despite these regulations, the river took a hand in their further ruination. The spring breakup of 1942 wiped the riverbank clean and the remainder of their wood was washed downstream. Needless to say, Laurent and Murray tried to avoid the wood business from that time on.

I have not personally checked the island for any habitat remains but you may wish to spend some time yourself poking around.

34 Excelsior Creek—Mile 343.5 (km 550)

MAP 12	PAGE 39

This is one of the two creeks by this name between here and Dawson City (see also mile 402). Martha Black, MP, a famous Yukon pioneer, named the creek in 1898. On her way to Dawson City, she stopped here for an overnight stay. She found some miners from New Zealand working the creek and they had not officially filed their claims or named the place. Overnight the group decided that the name Excelsior, suggested by Martha, won favor over the name Moari with which they were going to christen the creek.

The creek is only about six miles long and runs out of the mountains to the south. It is easily identified. It has a substantial flow and is one of the few creeks that flows directly into the main river. It has a distinctive rocky, sand and gravel makeup which is a rarity in this area. If the water level is from medium to low, check it out for a camping spot.

35 Ballarat Creek—Mile 344 (km 550)

MAP 12	PAGE 39

This creek was named by W. F. Woodward in 1898 after the famous gold fields in Australia. Over the years this locality has been occupied by homesteaders and miners almost continually. The creek is presently being mined and there should be signs of occupancy on the river bank. The last time I was there, in 1997, there was a large, white, fuel storage tank on the shore.

The creek itself is difficult to locate. It runs into the river some distance back and mixes with the Yukon River widely dispersed.

36 Coffee Creek—Mile 348 (km 557)

MAP 12   **PAGE 39**

This is one of the more substantial creeks in the area. There is generally a strong flow coming out from behind the island just upstream of the buildings shown on the map.

Coffee Creek has been occupied off and on since the turn of the century. The original settlement of Coffee Creek was spread out from the creek mouth and downstream for about half a mile to where the present landing is. For many years, pioneer Henry (Cy) Detroz had a farm at Coffee Creek. Henry arrived in the 1910s and lived there until the late 1940s. Detroz raised cattle and horses and other barnyard animals and generally grew as much as forty tons of hay annually.

There was a gold rush into Chisana, Alaska, in the early 1910s. A pack trail was constructed from here to Chisana and for a time Coffee Creek was used as a freighting terminus from where all supplies for Chisana were routed. Chisana is approximately 125 miles (200 km) southwest of here, across the border from present-day Beaver Creek, Yukon. Area residents Jack and Hazel Molloy ran a horse pack train into Chisana for a number of years and no doubt also used the farm as a source of feed and supplies.

During the Chisana rush a trading post came into being just down from the mouth of the creek. Although I show the remains of a building close to the creek, this has long since collapsed and disappeared. The original post was a small, single-story structure operated by Jim Derry. In the 1930s the post was taken over by the trading firm of Taylor and Drury out of Whitehorse and operated for them by Mrs. Hazel Molloy. The post did not remain open very long as there was very little business in the area. It closed in the mid-1930s.

Alternate routes were found into Chisana, and when the Alaska Highway was built in the early 1940s access to both Chisana and the headwaters of the White River became much easier.

Yukon character, Larry (Cowboy) Smith eventually gained title to the farm at Coffee Creek. Smith trapped the surrounding country for a number of years and raised horses at the farm. Like Detroz he grew most of his horse feed. Cowboy, originally from the Cariboo district of British Columbia, was more recently in the limelight as he entered both the Yukon Quest and the Iditarod sled dog races. The Quest goes for about 1,000 miles between Whitehorse, Yukon, and Fairbanks, Alaska, whereas the Iditarod pits dogmushers against each other between Nome and Anchorage, Alaska. Both are annual events.

In 1986, the farm was sold to Victoria and Kevin Olmstead of Fresno, California. They planned to develop the farm into a recreational facility. Plans did not go ahead and the farm was again sold in 1991 to U.S. citizens Rick and Deborah DeGraaf and their seven children. They had much difficulty with immigration officials in the early 1990s and I have not kept track of the goings on at Coffee Creek in the last three years. The place is occupied and should not be counted upon for an overnight stay.

37 Kirkman Creek—Mile 362 (km 579)

MAP 13   **PAGE 40**

This is named for Albert Kirkman who with an unnamed partner filed a discovery claim on the creek in 1898. The creek must have been unproductive as the original claims were allowed to lapse and a discovery claim was again filed in favor of Joseph C. Britton and William Heas in 1914.

In 1898–9 Francis Xavier Laderoute, a Gaspé Peninsula, French-Canadian, decided he recognized an opportunity when he saw one. He moved into Kirkman Creek, built a combination roadhouse, post office and residence cabin along the shore. He was hoping for enough business from the surrounding creeks to make a living.

On his way north and while he was in the midst of building Kirkman, Laderoute had stayed in touch with his family back east and particularly with his stepdaughter Marie Beaudoin of Grande Riviere, Quebec. In his infrequent letters to Quebec, he described Kirkman Creek as a virtual Mecca of the north of which he was mayor, postmaster, magistrate and mining recorder. He of course failed to mention that it was clearly 100 miles to Dawson City and also neglected to tell them about the long, isolated, dark and cold winters and the spartan living conditions. Perhaps Laderoute was genuinely unconcerned about the inconveniences and was one of those individuals that thrived in such surroundings. In any event, he obviously did not think that these minor details were worth mentioning. Laderoute's ramblings and the media attention given to Dawson City and the Klondike gold rush no doubt gave Marie and her family an attractive picture of the north. Marie fell on to hard times. During the first world war, her husband died of a sudden heart attack and Marie was left on her own with two infant children and only limited finances. Her thoughts turned to old stepdad and his glowing reports of Kirkman Creek. She possibly saw an opportunity to start anew. In the spring of 1918, she packed up her belongings and with the children in tow, headed north.

During the early hours of a June morning in 1918 the steamer dropped the gangplank at Kirkman Creek and Marie and her children arrived at Laderoute's domain. She was met by a small, wizened and goateed man who identified himself as Laderoute. As she stepped off the swaying gangplank, Marie took one look at her surroundings and realized that it was not at all what she had been expecting or had envisioned over the last eighteen years.

Laderoute had adopted some of the less desirable habits of a northern bachelor and the twenty long and isolated winters had also given him the typical introverted character which comes from talking to no one but the cabin walls too long. Marie, a normal excitable, vivacious, French-Canadian woman, found life at Kirkman very difficult.

Marie made a valiant effort to fit in but the visions of a long, cold and isolated winter overcame all other considerations and she decided to return to her native Quebec. The white flag was raised as the last southbound vessel came along and with a sigh of relief, Marie and her young family stepped on board, never again to return to Kirkman Creek.

Marie had some difficulty in expressing herself in English, which had of course not improved during her tenure at Kirkman. Laderoute had insisted on speaking French. When arriving in Whitehorse she was told that late season travel arrangements back to Quebec might be difficult. Marie was having some obvious problems in trying to explain where, why and how she wanted to leave the territory. Fate took a hand in the form of Antoine Cyr, a New Brunswick-born French-Canadian who arrived on the scene to lend a hand. Antoine must have done a remarkable job as five days later he had not only convinced Marie to stay but they were married. They remained in Whitehorse and raised a family of five more children. Antoine passed on in 1946 and Marie died in 1970 at age 77.

As for Laderoute, he remained at Kirkman until the fall of 1935, when he settled his affairs and left his river home for the first and last time. He boarded a steamer for Victoria, B.C. The change in climate obviously did not agree with him as shortly after his arrival there, he came down with pneumonia and died.

Marie, as his next of kin, was contacted for Laderoute's personal data which was not on his person at the time of his death. After much searching and writing back home, the parish priest in Quebec finally produced Laderoute's birth certificate. This showed his birth date as being 1828, making him a surprising 107 at the time of his death. What

Marie and Antoine (Tony) Cyr at Whitehorse in 1940s.   *Photo: Laurent Cyr.*

is even more surprising was that he participated in the Klondike gold rush, built his Kirkman Creek village and lived in the harsh north all after turning 70 years old.

Kirkman is renowned for its inherent gravel bars which made access and landing very difficult for the river steamers. To supplement his income, Laderoute had felled, cut and stacked a considerable pile of firewood on the riverbank, planning to sell it to passing steamers. The steamers' captains of course did not want to stop at Kirkman because of the navigational difficulties and Laderoute could only sell his wood if and when a steamer grounded. At this point the steamer would use up all her onboard supply of wood and would have to supplement its fuel supply from Laderoute's pile.

In 1935, Jack and Hazel Molloy moved into Kirkman from Coffee Creek. They apparently dealt with the estate of Laderoute on a purchase. It took several years for them to move their entire outfit out of Coffee Creek and it wasn't until 1940 that they could settle down completely.

The navigational problems in front of Kirkman give rise to another steamer story. Jack had taken sickly and the curing was beyond Hazel's ken. The accepted signal for the stopping of a river steamer was a white flag on the shore. If this signal was not there and the vessel had no cargo, mail or passengers for the place it would simply blow the whistle and keep going.

On the occasion of Jack's illness, Hazel had the flag up but the captain did not feel like dealing with the difficult landing. When it became obvious that the boat was not stopping, Hazel promptly raised her .30-30 rifle and put a round into the wheelhouse roof. She kept her rifle raised to indicate that the next one would be a little closer if she didn't see a change in course pronto. Jack got his ride into Dawson City!

Jack and Hazel lived at Kirkman until the mid-1970s. Jack died and Hazel being alone, decided out of necessity to move to Dawson City. Linda Taylor (nee Burian) and her family presently live at Kirkman Creek. Linda was brought up at Stewart Island about thirty miles downstream. It should not be counted on for an overnight stay.

The fairly new looking development on the downstream side of the creek belongs to Ed Kerklewich and family. It is not occupied on a permanent basis.

Limited mining is still done upstream on Kirkman and no doubt the history of this river village is far from over.

38 Carlisle Creek—Mile 364.5 (km 583)

MAP 13 PAGE 40

George Becker started a large wood camp here in 1935. He supplied both the river steamers and the city of Dawson with firewood and presumably rafted the firewood into Dawson. Becker died in 1936 and for the next three years Becker's foreman Alwyn "Taffy" Williams ran the camp for Becker's estate. Walter Isreal, (see Merrice Creek mile 241), bought the camp from the estate in 1939. Walter employed two families of Natives on a permanent basis, moved in some horses to do the heavy hauling and the camp was a beehive of activity. Isreal could recall one winter cutting a record 755 cords of wood. His primary customer was the White Pass & Yukon Route which operated almost all of the steamers on the river at that time.

Walter sold the camp in 1946 to Ragnar Nelson and Bud Holbrook They only operated the camp until 1947 and abandoned the site at that time. No signs of occupancy remain. Isreal reclaimed the site no doubt for debts owing. He sold the buildings at the site. They were dismantled and moved.

Carlisle is the name of a city in northwest England. We have no idea who named the creek or when but assume that it was named for this city, perhaps by a homesick miner.

39 Touleary Creek—Mile 351 (km 561)
Los Angeles Creek—Mile 369 (km 590)

MAP 12, 14 PAGE 39, 41

Both creeks were presumably named by Albert Kirkman (see Kirkman Creek mile 362). Kirkman was from Tulare, California. Kirkman and his crew filed claims on both of these creeks. We assume that over the years Tulare became Touleary. The crew being from California obviously named Los Angeles Creek for that city in the United States.

40 Thistle Creek—Mile 370 (km 592)

MAP 14 PAGE 41

A group of eight Scotsmen settled in and around this stretch of river in 1898. We assume that they had been to Dawson City and like most late stampeders found the town crowded, expensive and disheartening insofar as the likelihood of staking a claim in the Klondike was concerned. The number of people in Dawson City at that time made even the most simple supplies hard to come by.

Thistle Creek roadhouse in 1975.   *Photo: Gus Karpes.*

Couture cabin at Thistle Creek, 1996.   *Photo: Gus Karpes.*

Two of the group, Murdock McIver and Robert Haddow, liked what they found at this creek and filed the discovery claim on September 28, 1898. The claim was about six miles above the mouth of the stream. They named the creek after the Scottish national emblem. They explored the countryside and in the course of their wandering named Scotch Gulch, Green Gulch and Thistle Mountain.

A small settlement sprang up near the creek which they called Thistleton. This of course also called for the inevitable roadhouse or gathering place at the site. These roadhouses or guest houses materialized along the river at every location that so much as hinted of a settlement. No doubt they could be compared to today's community halls or highway lodges, a place for travelers and locals alike to meet, eat, drink and socialize away from the solitary, single-room cabins that most of them lived in. The roadhouse sat on a hill behind the cabin that is there now. It was taken down in the early 1980s.

The building that still stands at the site was last occupied by the Couture family. Gerry and his wife Jan raised three children there in the 1980s. They now reside in Dawson City and the site has been unoccupied for some years. It is still private property and should be left alone.

If you plan on landing, be careful and land well upstream of the site. The bank in front of the residence erodes continually and the current is very swift.

The creek itself is difficult to find and enters the river in a back slough. The creek is currently being mined and some distance downstream you will be able to see the large fuel-storage tanks located at the mouth of the creek. Modern mining methods and the price of gold have reactivated mining at many of the creeks between here and Dawson.

41 White River—Mile 380 (km 608)

MAP 14   **PAGE 42**

There is no mistake as to why this river is named the White. Its heavy, silt-laden water comes into the main Yukon just ahead of the Stewart River. Except for its upper reaches, the whole length of the White is one mass of confused and disorderly channels, generously complemented with an array of backwater sloughs and blind alleys.

The river was named by Robert Campbell in 1851. Schwatka indicates that the local Natives called it the Sand River.

The White originates in the St. Elias Mountains to the west. This mountain range has some of the most glaciated peaks on the North American continent. The river's color is caused by an eroding, heavy layer of volcanic ash in the entire White River Valley. This was deposited by volcanic eruptions in the St. Elias Range. Some great eruptions took place around 900 to 1000 A.D. and as recent as the eruption of Mount Katmai in 1912 which spread a white blanket of ash throughout the whole region. In places the ash layer is said to be almost 100 feet thick.

You can now access to the headwaters of the White River from the Alaska Highway close to the Alaska–Yukon border at Beaver Creek. I know of a few people who have done the trip from there to Dawson City by canoe. None describe it as a very exciting trip, mainly due to the quality of the water and the difficulty in navigating the river.

Since the first exchanges of information between the white men and the northern Natives, it was rumored that a very rich deposit of copper could be found in the White River Valley. This rumor gained substance when a Native at Fort Yukon, at the mouth of the Porcupine River, showed off an almost pure copper piece of bauble and very reluctantly parted with the information that it came from the White River Valley.

Arthur Harper, a very early Yukon pioneer, and three companions spent the winter of 1873–4 in the White River Valley unsuccessfully pursuing this elusive ore. Harper also panned for placer gold but nothing of consequence was found.

William Ogilvie attempted to navigate upstream on the White in 1887 but was turned back as the water totally confused his boat men.

Frederick Schwatka, chasing further personal fame, returned to the Yukon River Valley in 1891 and mounted an expedition into the White River Valley. The results of his efforts were never published as he died shortly after his return to the United States.

For a time, the excitement of the Klondike gold rush overshadowed everything else, and the White River and its fable of rich copper deposits was forgotten—but not for long. In 1905, Solomon Albert found a promising copper deposit about 130 miles upstream on the White and about ten miles from the Yukon–Alaska boundary. The settlement of Canyon City was founded and became the "head of navigation" on the White River. Personally we cannot visualize anyone navigating the White River, but obviously some of them did.

With Canyon City as a base, it wasn't long before prospectors and exploration parties went further afield. In the early 1910s the Chisana River and the Wrangell Mountains became the source of great excitement as gold in paying quantities was found. The supply demands of Canyon City and Chisana were high and for a brief time the residents of Coffee, Kirkman and Thistle creeks were busily engaged in the operation of a pack-train service into the area.

This flurry of commerce was short lived. The Chisana Trail was built from the Alaska coast into the Chisana country. The copper deposit at Canyon City did not prove to be a major find and warranted no further development. As the mystery of the White River copper had been solved and Chisana had its own trail in from the coast, the White River Valley was abandoned as a transportation corridor and there has been little or no further activity in the valley since.

One of the more interesting tales to come out of the White River Valley involves Solomon Albert, the founder of Canyon City. While poking around the headwaters of the White River, Solomon ran into difficulty of sorts which resulted in him getting his feet wet. They quickly froze in the frigid winter temperature. He managed to stumble into J. Slaggard's camp near Canyon City, then abandoned. Slaggard brought Solomon to within thirty miles of help but the party could not continue. They had no food left for themselves or their dogs and could no longer carry Solomon on the sled. Slaggard had to leave Solomon on the trail. He started a fire, left some fuel and mushed on to the Yukon River to get help or at least send a rescuer back for Solomon. One husky stayed with the injured man. This was Solomon's last dog and it refused to leave him. Slaggard and his dogs made it to a food cache on the Yukon River belonging to Billy Roup. Roup and Slaggard with several horses went after Albert. The trip was horrendous, with the two at one point having to break trail for the horses through open water. This meant wading into the water up to their hips and once through, having to stop, light a fire and dry themselves out. They finally came upon Solomon Albert who had had to kill his faithful dog to survive. They started back and were met by a mounted police patrolman before they came to the Yukon River. Solomon was transferred to his sled and taken to Dawson City.

All of his toes were amputated. This made walking difficult. Later, when Solomon got better, he apparently killed a grizzly bear, carefully skinned out its paws and slipped two of them over his footwear. The bear claws would act as toes and he was able to keep his balance while walking. For years he was known as the "Bear Man" and he would go stomping about Dawson City with the makeshift footwear tied on. People could hear him coming from far away as the claws would scrape on the wooden boardwalks of the town.

42 Stewart River/Stewart Island
—Mile 390 (km 624)

MAP 15, 16   PAGE 42, 43

The Stewart River was named by Campbell in 1851 for his assistant James Green Stewart. Stewart was with Campbell during the Fort Selkirk days and was the man Campbell left in charge of the fort when he floated from the Pelly to Fort Yukon at the Porcupine River confluence. Stewart was no doubt responsible for much of the construction of the fort.

As early as 1883, the year of Schwatka's trip down, four men were known to have been on the Stewart River for the entire summer. Leroy McQuesten met the men in Tanana, Alaska, during that winter and described the meeting with the four in his memoirs. The men claimed that they made $10 per day rocking the bars of the Stewart and some of its tributary creeks—a very healthy sum in those days.

The winter of 1885 that McQuesten spent at Stewart was the coldest he had encountered in the north. On the fourth of January, 1885, the temperature dipped to -80 degrees Fahrenheit and the average temperature for that month was -57 degrees.

In 1886, as many as 100 miners were panning the river bars on the Stewart River, some of them quite successfully. At that time the Stewart offered the best "color" found anywhere in the district. Also in 1886, Arthur Harper built a trading post at the mouth of the Stewart and since that time the location has rarely been without some permanent activity. For many years, into the 1940s, the Stewart River was the main transportation artery to and from the silver mines in the Mayo, Keno and Elsa districts. Smaller boats were used to run up and down the shallow Stewart River. These boats were met at Stewart by the larger vessels running the Yukon River. The ore from the mines was transferred to the larger boats and barges for transport to Whitehorse.

One of the boats used on the Stewart River was the SS *Keno*. This is now docked in Dawson City as a museum. It is worth a visit.

Once the Klondike Highway was built into Mayo and Dawson City, Stewart City fell into total disrepair and consequent development has taken place at Stewart Island just downstream. This island was often referred to as Scow Island as a large number of scows and barges were stored and later abandoned in the slough behind the island.

Stewart Island is the home of the Burian family. For many years,

Rudy and Yvonne Burian welcomed river travelers to Stewart Island, a virtual oasis is the wilderness. They maintained some rental cabins, a store and the Stewart City museum. Rudy has passed on and Yvonne is badly incapacitated by arthritis and now lives in Dawson City. To make matters worse, Stewart Island has been gradually disappearing into the river. The following photographs will give you an idea as to how much of it has disappeared. In the 1975 photo there were literally acres of grass and garden. Most of that has gone into the river in the photo taken in 1996, and in the spring of 1997 the blue house was undermined and also slid into the river.

Robin Burian still lives at Stewart Island and maintains his cabin there. It should not be counted upon for an overnight stay or for the purchase of supplies.

Stewart Island, 1975.   Photo: Gus Karpes.

Stewart Island, 1996.   Photo: Gus Karpes.

Yvonne Burian (center) at Stewart Island, 1993.   *Photo: Harry Kern.*

43 Sixty-Mile River—Mile 413 (km 661)

MAP 18   **PAGE 45**

This was named by Arthur Harper for its distance from old Fort
Reliance which was about eight miles below present-day Dawson City.
The trading post at Fort Reliance was the major trading center from
1874–86. It was then abandoned in favor of a post at the mouth of the
Stewart River.

Harper filed the original claims on the Sixty-Mile River in 1876.
In 1891, its tributaries showed some good results which started a minor
gold rush into the area.

The Sixty-Mile Trading Post was opened by Joseph Ladue and
Harper in 1894 on the island just opposite the mouth of the river.
Yukon's first sawmill was constructed just downstream. Shortly after
the trading post was opened, it was renamed Ogilvie for William
Ogilvie, DLS of the Department of the Interior. Ogilvie was a promi-
nent government representative in the territory.

With the Klondike discovery, the post lost its prominence in the
local community affairs. The sawmill was moved to Dawson City and

the island site was abandoned. Very little evidence of this once-bustling community remains.

The Sixty-Mile tributaries still yield their share of gold. The river is now accessible via the Top of the World Highway out of Dawson City. Just downstream of the river you will see a cleared area with a derelict tank and equipment sitting on the shoreline. I believe this site has not been used for years.

44 Indian River—Mile 434 (km 694)

MAP 19	PAGE 46

Probably named for an early river traveler for an Indian settlement located at the mouth of the river. There are two Indian Rivers in the Yukon Territory, the other runs into the Teslin River. The range between Indian River and the Klondike River enclose all of the rich placer creeks in the Klondike region.

45 Ancient Voices Wilderness Camp —Mile 434 (km 694)

MAP 19	PAGE 46

What we have lost at Stewart Island, we have gained here. Peter and Marge Kormendy, Bertha Blondin and James Babineau have created a wilderness camp at Galena or Dog Creek. The four of them represent a Native heritage from Yukon, Northwest Territories and Alaska. The

Ancient Voices Wilderness Camp, 1996.   *Photo: Marge Kormendy.*

camp is geared to river travelers and from all accounts it is a marvelous place to spend a couple of hours to several days or more.

On a pay as you go basis, you can participate in any number of traditional Native activities from tanning hides to smoking fish. You can use your own tent or give yourself a break and spend the night in a wall tent or log cabin from the variety that they have to offer.

46 Swede Creek—Mile 452 (km 723)

MAP 21   **PAGE 48**

The first discovery claim was staked here on February 2, 1898, by C. A. Olafson and his Swedish crew. The initial strike resulted in another "rush" to the creek and the staking of more than 600 claims.

The creek was named for the nationality of Olafson and his crew presumably by the mining recorder in Dawson City.

47 Klondike River—Mile 460 (km 736)

MAP 21   **PAGE 48**

The name Klondike stems from an early Indian name of Thron-Diuck or Tron-Deg depending on whose interpretation we accept. The "T" is usually pronounced as a "K." These words, literally translated, we believe meant "Hammer Water," derived from the Indian practice of driving stakes into the bed of a stream to form fish traps for the migrating salmon.

After much tongue twisting with the original name, the miners finally settled on the pronunciation of "Clunedik" or "Clundyke." Inspector Charles Constantine of the Northwest Mounted Police, then also acting as the mining recorder for the district, officially adopted the name "Klondyke." Usage over the years substituted an "i" for the "y" and consequently Klondyke became Klondike. The first report of the Canadian Board on Geographical names made the name "Klondike" official in 1898.

48 Dawson City—Mile 460 (km 736)

MAP 21   **PAGE 48**

Dawson City is named for Dr. George Mercer Dawson (1849–1901), Director of the Geological and Natural History Survey of Canada from 1895 to 1901.

I will not attempt to provide a history of Dawson City and the Klondike as there are many fine books available that do an admirable job of providing this history.

6

River Tales

The *New Racket*

The *New Racket* was a little steamer that first appeared on the lower Yukon River in 1882. Edward Lawrence Schieffelin, the discoverer of Tombstone, Arizona, built the small vessel in San Francisco. Schieffelin describes her as a "small sternwheel steamer, of about fifteen tons burthen." He and his party shipped the small steamer to St. Michael's, Alaska, on board a larger schooner which they had chartered.

Schieffelin and his party remained in Alaska until the fall of 1883. He sold the *New Racket* to the pioneer team of McQuesten, Harper and Mayo. Stewart Menzies, a pioneer, for many years the auditor for the Alaska Commercial Company, wrote historian Clarence L. Andrews on an incident involving the *New Racket*. The letter is dated April 14, 1918. Menzies writes:

"In the fall of 1895, I took the Str. Beaver up the Yukon from St. Michael, arriving at Pelly on the 13th of October. The nearest safe wintering place was a slough about four miles above the post. In the slough, pulled up on the bank was the remains of a little stern wheel steamer which I afterward learned was the famous New Racket which was originally brought into the country by Ed Sheffelin somewhere about 1882 or 1883."

During the winter, Arthur Harper who had the trading post at Fort Selkirk where Menzies wintered, had the idea of floating the *New Racket*. Just before break up in 1896, Menzies set about making preparations for the launch. I continue Mr. Menzies' tale:

"I had just got the boat clear of the water when the main Yukon ice broke and jammed on an island about three miles below the post. This jamming of the ice naturally caused the water to back up very rapidly, in point of fact the slough rose about 20 feet inside of five minutes, lifted Mister New Racket up and landed it back in the timber about a quarter of a mile where it landed gracefully on top of a stump which penetrated the bottom of the boat, and as far as I know she is there yet."

From Menzies' description, the slough can only be one of the backwater branches of Slaughterhouse Slough. On a number of occasions I have thought of launching a search for the remains of the little vessel but have never had the time to do so. Perhaps parts of the little steamer are still there.

One other remark of Mr. Menzies' is interesting as he describes the havoc that the ice jam caused.

"At the same time the water on the flat at the post was about four feet deep and Harper had a fine time drying out such of his stock of calicoes, drillings, etc., as had not been put on high shelves in the store."

The post that he refers to was at the site of current-day Fort Selkirk.

Pelly River Pioneers

The Pelly River Valley has always been very bountiful in its timber. Around the turn of the century, building materials and firewood became scarce around the settlements in the Yukon River Valley, particularly around the city of Dawson. Imagine the timber used in the early days of Dawson City simply to build the town, the satellite communities and the mines. The largest consumption around the communities would be for firewood, not only to keep everyone warm during the cold winters but to also feed the cook stoves in summer that were to feed the thousands of people that suddenly descended on the Klondike.

More than 250 sternwheel steamboats operated on the rivers during the busy years. Each could burn between one and three cords per hour. This would depend on the river, the load and the number of barges they were pushing. It is hard to even estimate the amount of wood cut throughout the river valleys just to keep the boilers going in these vessels from year to year.

Pelly River and Fort Selkirk residents supplemented their income by logging the Pelly River Valley. They made up huge rafts and from there, rafted the wood downriver to Dawson City. Two of the following photos show Little Sam's log rafts tied up at Fort Selkirk. The raft shown in the photos was made up of 110 cords of wood. The third photo is of Little Sam and his crew.

94

LiTTLE. SAM . FoRT SELKiRK.
YT
1947

Top and Center: Log rafts at Fort Selkirk.   *Photos: Wm. L. Drury.*
Bottom: (left to right) Edward Simon, Little Sam, David Silas and Frank Blanchard.

The Pelly River farm, one of the earliest farms in the Yukon Territory is located about four to five miles upstream from the confluence of the two rivers. The 337-acre (136.5 ha) farm was first staked out in 1901 by Edward Menard, a telegraph operator at Fort Selkirk. The property went through a number of owners including Frank Fairclough and his family in the 1920s. Frank supplemented the farm income by rafting white spruce timber to Dawson City. There were owners Chapman and Olsen who apparently raised cattle and grain. Later owners were the J. C. Wilkinsons who used the farm as a base camp for their outfitting business in the 1940s and early 1950s. In 1953, John Stelfox of Rocky Mountain House, Alberta, with brothers Hugh and Dick Bradley and Buck Godwin all of Olds, Alberta, purchased the Wilkinson farm. John Stelfox left the Yukon in 1955 but the Bradleys remained and are still involved in the running of the ranch.

In the early years the farm and the Pelly roadhouse, situated where the winter road crossed the Pelly, formed their own little community most of which has long since disappeared.

Native Dress in the 1800s

The deer is a recent immigrant into the territory and it is generally believed that they migrated into the Yukon during the last fifty years. There are some unconfirmed reports of deer being present in the southeastern quadrant in the early 1900s. There are also reports of deer sign while the Alaska Highway was under construction in the 1940s, but there were no confirmed sightings of the animal.

In his description of the two leading chiefs that he met in 1843, Campbell indicates that they were clad head to foot in dressed deer skins. The skins that the Indians dressed in were in all likelihood caribou hides. The Forty-Mile caribou herd at that time was estimated to number around 200,000 animals and their migration route took them as far south as Carmacks.

Taylor and Drury—Pioneer Merchants

Taylor and Drury Ltd. was a turn-of-the-century trading firm started by Isaac Taylor and William S. Drury. They operated the trading post at Fort Selkirk. The large log building sits very prominently on the river bank—you can't miss it. The firm also had stores and trading posts at Whitehorse, Carmacks, Minto, Little Salmon, Coffee Creek, Teslin and Ross River.

The company boat, usually pushing a barge, made two trips a year to their various posts. The village of Teslin was reached by traveling upstream on the Teslin River from Hootalinqua. The village of Ross River was reached by traveling about 260 miles (416 km) up the Pelly from here. The *Yukon Rose* and her barge are shown in the following photo as they are about to leave Whitehorse for Ross River in 1931.

Yukon Rose and barge leaving Whitehorse, 1931.   Photo: Wm. L. Drury.

On August 23, 1933, the company vessel *Yukon Rose* was tied up at Fort Selkirk. On board was Drury's fifteen-year-old son, Thomas E. Drury. Young Tom apparently tripped and went overboard. He was a more-than-adequate swimmer and waved to those on deck as he drifted downstream. Suddenly, apparently overcome by cold or current, Tom slipped beneath the surface of the water and did not come up again.

William (Bill) L. Drury, a younger brother, related the incident and said from that day on he was not allowed to go with his father on any of the boat trips and did not do so until well into his late teens. In memory of Thomas Drury, the family donated a window to St. Andrew's Church. Walk to the back of the church and you can see this stained glass window which is still in remarkable condition. The right, upper pane carries the initial D for the Drury family.

97

St. Andrew's Church, Fort Selkirk.   *Photos: Gus Karpes.*

Window detail, St. Andrew's Church, Fort Selkirk.

Industry along the River

The wood business was a big business during winter in the Yukon. Each river steamer burned as much as 160 cords of wood on a return trip between Whitehorse and Dawson City, forty cords on the downstream run and the rest on the upstream part of the trip. A vessel could make as many as ten or twelve trips per season.

Each good-sized vessel carried a crew of about thirty-five. Ten would be officers down to the first mate and engineer. The rest of the crew would consist of deckhands, cooks, stewards and engineroom people. Around the 1930s and the 1940s the wages for these crews ran as low as $60 per month for the deckhand and dining-room stewards to as high as $150 per month for the deck and engineroom officers. The captain, pilot and chief engineer were the highest paid people on board, with a salary of around $300 per month. The boating season generally ran from May to the first part of October.

The boats usually ran twenty-four hours a day and would use as little as twelve and as much as thirty cords of wood in a twenty-four-hour period. This would depend largely on what the ship was doing. For instance if it was going upstream, pushing a number of barges, it would probably burn around three cords per hour; if going downstream, it was more like one cord an hour. If it was stuck on a sand bar and the winches, paddlewheel and other onboard equipment had to be used, it could burn a considerable pile of fuel just getting off.

A lot of the crews lived outside and came into the Yukon for the summers only. A great number of them also lived in the Yukon. Cash was scarce and the wages that a deckhand or cook made on the boat was a welcome part of the local economy. Quite often the summer crews turned into woodcutters in the winter time. This was so common that when an alternate fuel such as oil or coal was considered, this winter employment of the crews was a deciding factor in rejecting the idea. The company reasoned that if there was no employment for the crews in winter, most of them would leave the country and they would have to train a new crew each spring.

The Makings of a Discovery Claim

A discovery claim filed under the Placer Mining Act is the first claim registered on the creek. Anyone filing a discovery claim named the creek, if it did not already have a name, and pinpointed its location to the mining recorder. A discovery claim was 1,500 feet. Consequent claims were numbered from the discovery claim and were only 500 feet each. Number four above discovery would therefore be the fourth 500-foot claim upstream of the discovery claim. If the original claims along a creek were allowed to lapse, a discovery claim could again be filed.

The Hunting Party

"What a way to go. Look at this! On-board accommodation yet. And look at that stove will ya!? This is what I call a first-class hunting trip."

The speaker, all 300 pounds of him, we affectionately called Tiny. He was in his sixties, about six-foot-seven in his stockinged feet and could just barely stand up straight in the boat cabin.

"Yeah, this sure beats the last trip I was on." This time from Raymond. He had recently moved to the Yukon and was full of enthusiasm for anything new. In his late twenties, he was the youngster of the crew. As we were on a moose hunting expedition he had also been elected as the surgeon or meat cutter. He had successfully made his way through medical school and we naturally assumed he would be up on his cutting procedures.

The fourth member of our party was doing what he loved best. He had the engine hatches open and his head stuck into the engine compartment. The Pooh, as he was called, was a mechanic by trade and loved his work. His given name was Winston, which was almost immediately shortened to Win and the Pooh came naturally after that.

The boat was thirty-six feet long, seated twelve and in the summer did duty as a tour boat out of Whitehorse. It was one of the two boats that Irene and I operated between Whitehorse and Dawson City on the Yukon River in the early 1980s. The seats came down for bunks and, at least on this trip, there would be no need to pitch a tent. The four of us were out to get the winter's meat in the form of a moose or several moose for that matter. All of us had a license and a tag. What we were collectively spending on this one-week hunting trip would have probably bought our meat for the rest of the winter. What the hey! It would not be near as much fun and would not really do much to satisfy the age-old primeval urges of man the hunter and provider.

We were busily engaged in bringing the provisions on board. Norman had been put in charge of the victualing as he liked to call it, and believe me, he was a man who loved his food. Being from the east coast of Canada he also liked to impress on everyone his knowledge of nautical terms. His first line set the stage for the rest of the trip.

"Do you want the victuals in the port or starboard lockers skipper?"

"Put them in the foredeck locker please Norman. They'll stay a little cooler up there," I replied casually.

"Sheeit, Pooh! We're in the company of a couple of real old salts. You and I are going to be relegated to the kitchen sure as shootin," Raymond complained. "Should have brushed up on the lingo before I left."

"Galley, you cretin. On board ship you call it the galley," Norman added for good measure.

Soon everything was stored in its proper place and it was take off time. Our destination was Carmacks, about 200 miles downstream. We had all taken seven days off and it promised to be a great trip. It was the middle of September, a little cool to be out in a boat but with all the comforts of home around us, that was the least of our worries.

We had two rifles on board, an old .303 that Norman brought and which had gone through most of the last war with him, and a .270 Winchester which was mine. The other two did not own a gun, but that had not stopped them from grabbing a license and jumping aboard.

"Let's do it. Ray, you're the youngest and the deckhand. Win, you're the engineer, Norman is the cook and I'll try and get us there in one piece."

I started the engine and gave it five minutes to warm up.

"Okay Ray, let go the lines. Stern first. That's the one at the back of the boat."

We were off and soon Whitehorse was behind us and we settled down to enjoy the trip. It was a beautiful September day—blue sky, fall colors and just a skiff of snow on the mountain tops.

Day five and we were into a schedule. Up with the sun, one of Tiny's gourmet breakfasts, push off and cruise till lunch. We had had grayling for breakfast a few times, but didn't have a hint of moose.

"Weather is too nice," remarked Ray. "Maybe we should try getting up in the morning."

"Nah!" Norman joked, "Moose like to have breakfast too. We're just a little too early in September that's all. We'll get one," he added confidently.

One of us rode on the upper deck as the look out. Pooh was on shift as we approached Dutch Bluff, a prominent 200-foot-high sand and clay bluff that you can see for miles around. The constantly sliding sand makes picking a channel a bit of a challenge, so I was busy trying to keep us out of trouble.

"Moose in the slough to starboard," came a shout from above.

After five days of being subjected to Norman's tutoring, everyone knows where starboard is. There was a great amount of scrambling going on. I managed to take a look and glimpse a large bull moose standing in the slough behind us.

"I'm going to come around you guys. Norman, grab a rifle and get ready to take a poke at him."

I swung the boat and headed toward the point of the island. I have always found that if engine revs are kept fairly steady and no sudden moves are made, the animal doesn't seem to spook and their natural curiosity keeps them standing there. As I swung the boat, the moose stood head-on about 100 yards away. Perfect! Next thing you know 300 pounds of Newfie hits the deck and in perfect rifle range fashion, presents arms and lets fly.

"Missed, dammit." Norman flipped back the bolt, the spent casing did a swan dive and we were ready for a second shot.

The noise of the shot spooked the bull. He decided that maybe it wasn't all that healthy for him to keep standing there. As the boat beached at the head of the island, he took off into the willows which were so thick that he was out of sight almost immediately.

"Okay guys, he's on the island. Grab the rifles and let's go. It's a small island and he can't get very far without us seeing him," I told the crew.

I turned the engine off, took a quick wrap around a large willow and off we went. Pooh and I went one way, Norm and Raymond the other. Winston carried my gun.

"Watch where you're shooting," I cautioned them. "We don't want to shoot each other."

Pooh and I were about fifty yards into the bush of the island when we stopped to listen. When a large moose with a rack upwards of six feet travels through the willows, you can hear him. There is no way that he can keep himself quiet with that big coat rack on his head.

"Click" and then again "click, click." Suddenly the tempo picked up until it sounded like someone rattling a stick along a picket fence.

"He's heading for the beach," I shouted and started running to my left. There were willows, dead falls, rocks and clumps of what not in the way, but I made good time and hit the beach about the same time that the moose skidded to a halt about fifteen yards away. I stopped, huffing and puffing from the effort. He too stopped, turned to face me and let out a loud, nose-shaking snort.

"Quick, give me the gun," I shouted and turned around to reach for the piece. No one there! I glanced back and tried to keep my eye on the moose at the same time. Back about twenty or so yards into the willows I could see Pooh struggling to get up. I turned my full attention to the moose. I swear I could count the pimply hairs on his nose and he was looking at me with what I can only describe as a deignful look, looking down his nose as much as to say, "What now dummy?" He stamped his front hooves for emphasis.

Now, I don't know whether you've ever been that close to a large wild animal before. But I knew right away that I was just a little outclassed without my fowling piece. It flashed through my mind that an old friend, when confronted by a bear in much the same circumstances, took his hat off and started stomping it into the ground, all the while yelling evil incantations to do with the animal's parentage. The bear had apparently grunted once, turned around, dropped down to all fours and shuffled off, totally intimidated. "Should work for a moose," I muttered.

I took off my hat, threw it on the ground and was getting ready to do a war dance. I picked up a rock for insurance purposes and fired it off in the general direction of his nose. I missed and was just about to start stomping my good hat into the mud, when Pooh scrambled out of the willows.

"Gimme the gun Win! Gimme the gun! Hurry up, he's going to get away," I yelled at him.

Pooh's sudden appearance confused Mr. Moose. He reappraised his position and decided to leave for greener and less-threatening pastures. With a startled snort, he turned and started legging it out of there. While I was still trying to get the artillery organized, he splashed across to the next island, ducked around a tree or two and by the time I shouldered the gun, he was well over 100 yards away and I was looking at his hind end. Disgusted, I let fly with one anyway. Of course I missed, and didn't get a second chance before he disappeared around the bend.

We never did see another one and that single incident was our only wildlife encounter of the trip. We covered 200 miles of river unscathed and despite the lack of success, it was a memorable trip. The fearsome four managed to stay out of further trouble until the end.

The story of the confrontation with the moose was repeated and embellished in the telling a time or two as most hunting stories are. I am sure the tale is still a favorite around a few northern camp fires.

Occasionally, during a fitful sleep on a long winter's night, I can distinctly see those two beady eyes looking down at me from behind that upraised nose. To keep it from turning into a nightmare, old Winnie always shows up in time and I miss getting gored. Sadly, I never get to find out how effective my war dance would have been. Ah well...at least I saved my favorite hat.

Sanctuary

It has been many years since I took my first trip on the Yukon River. I remember the trip vividly. It made such an impression on me that since then, the river has played (and will continue to) a great part in my day-to-day life.

I have traveled the Yukon River in a variety of watercraft. This includes an assortment of canoes and power boats, as well as a bathtub in what is billed as the longest bathtub race in the world. I sailed as the captain and pilot of the riverboat MV *Anna Maria* in the late 1980s. She is likely to be the last large vessel that will ever operate on the Upper Yukon River between Whitehorse and Dawson City.

Irrespective of how often and in what manner of craft I journey forth, I still experience an almost euphoric feeling of anticipation every time I set out on a voyage. To my mind, nothing equals the experience of feeling that first tug of the river current on the canoe or boat, knowing that for the next few hundred miles and a number of days or weeks, I will be totally involved with the river. My daily life will be regulated by it. I will become one with nature and be allowed to view firsthand its many wonders. Most of all, I look forward to being able to do this totally on my own. I will be dependent on my own resources and those of my companion or companions. There are no yellow lines dividing the two sides of the river. There are no traffic signs, no stop here, no loitering, turn here signs regulating my daily progress. I can eat when I want, stop when I want, pitch my tent where I want, and travel at my own pace, whatever it may be.

For a number of years, my wife and I were involved in river outfitting. During those years, it was apparent that this almost unheard of freedom of movement and existence, was one of the more frightening expectations for the would-be, first-time river travelers. Unable to handle this unregulated lifestyle, some stopped short of their intended destination, others turned the trip into a marathon in order to shorten their time alone and still others feigned ill health or bad weather so as to justify their early abandonment of a trip. Quite a number of them admitted they were totally out of their element and could we please tell them where the nearest shopping mall was located. So you see an extended wilderness river trip is not for everybody.

Frequently my advice to a solo canoeist or an inquisitive couple was that perhaps their greatest experience of the trip would be learning to live with nature, its prescribed boundaries, its limitations and learning to live with themselves or their single companion within these confines. Rarely in today's worldly environment do we spend our total day time alone. Rarely do we spend twenty-four hours, day after day, with a single companion who is virtually never out of sight. Entire books, magazines and classroom courses are devoted to teaching people how to live, work and play together as couples, families or workplace companions. A trip such as this can be as informative and rewarding as the best of them.

An extended wilderness river trip, or for that matter any kind of an unfettered wilderness adventure, can be one of the most rewarding experiences in life. An open fire, today normally associated with injury, loss or threat, becomes a comforting presence at the evening campsite. The companionable spell it casts over most of us, gives us that primitive and basic sense of well being. We become intensely aware of the world around us. The birds, the trees, the insects, the wind and the river become part of our everyday existence. We learn to identify and appreciate the meaning and makeup of our natural surroundings. We learn about ourselves and our companions. I feel that if we openly accept all of this input during our venture into the wilds, brief as it may be, we cannot help but come out of the experience a more complete and better person.

I hope that this guidebook has been useful and interesting and that the trip itself has fulfilled your expectations.

References

Berton, Pierre. *Klondike: The Last Great Gold Rush*, McLelland and Stewart, Toronto, Canada, 1985.

Burian, Mrs. Yvonne. Personal interviews, Dawson City, Yukon and Stewart Island, Yukon.

Carmack, George W. *My Experiences in the Yukon*, The Trade Printery, date unknown.

Coutts, R. *Yukon Places and Names*, Gray's Publishing Limited,1980.

Cyr, Laurent. Personal interview, Whitehorse, Yukon.

Dawson, George M. *Report on an Exploration in the Yukon District, N.W.T. and Adjacent Northern Portion of British Columbia - 1887*, Yukon Historical and Museum Association, 1987.

Drury, William L. Personal interviews, Whitehorse, Yukon.

Gairns, D. D. *Scroggie, Barker, Thistle and Kirkman Creeks, Yukon Territory Canda Department of Mines—Geological Survey Memoir 97*, Ottawa Printing Bureau, 1917.

Hamilton, W. R. *The Yukon Story*, Mitchell Press Limited, 1964.

Hoefs, Manfred. Personal interview, Dept. of Renewable Resources, Government of Yukon,1997.

Isreal, Mr. & Mrs. Personal interview, Carmacks, Yukon, 1979.

Johnson, James Albert. *Carmack of the Klondike*, Epicenter Press, Horsdal & Schubart Publishers Ltd., 1990.

Maclean, Hugh D. *Yukon Lady*, Hancock House Publishers, Surrey, British Columbia, 1985.

Martinsen,Ella Lung. *Black Sand and Gold*, Binford & Mort, 1974.

Menzies, Stewart. "Some Notes on the Yukon," *The Pacific Northwest Quarterly*, Volume XXXII, The University of Washington, 1941.

Recollections of Leroy N. McQuesten, Life in the Yukon 1871 - 1885,
Dawson City, Canada, Dawson Lodge No.1 Yukon Order of
Pioneers, 1977.

Sawatsky, Don. *Ghost Town Trails of the Yukon,* Stagecoach
Publishing Co. Ltd., 1975.

Schieffelin, Edward Lawrence. *History of the Discovery of Tombstone,*
Arizona, Red Marie's, Tombstone, Arizona, 1988.

Schwatka, Frederick. *A Summer in Alaska,* J. W. Henry, St. Louis, Mo,
1894.

Shand, Ora and Margaret Clark Shand. *The Summit and Beyond,*
Caxton Printers Ltd., 1959.

Smith, Delores. "Pelly Farm served as the Wilkinsons' trapping base,"
The Whitehorse Star, January 17, 1996.

Journals of Robert Campbell (Chief Factor Hudson's Bay Company—
1808 to 1853). Typed and marked 'limited edition,' Seattle,
Washington, 1958. Vancouver Public Library - 61-1781-10.

Wilson, Clifford. *Campbell of the Yukon,* Macmillan Company of
Canada Ltd., Toronto, 1970.

Wright, Allen A. *Prelude to Bonanza: The Discovery and Exploration*
of the Yukon, Gray's Publishing Ltd., 1976.

Index